USING SCIENCE
AS EVIDENCE
IN PUBLIC POLICY

Committee on the Use of Social Science
Knowledge in Public Policy

Kenneth Prewitt, Thomas A. Schwandt, and Miron L. Straf, *Editors*

Division of Behavioral and Social Sciences and Education

NATIONAL RESEARCH COUNCIL
OF THE NATIONAL ACADEMIES

THE NATIONAL ACADEMIES PRESS
Washington, D.C.
www.nap.edu

THE NATIONAL ACADEMIES PRESS 500 Fifth Street, NW Washington, DC 20001

NOTICE: The project that is the subject of this report was approved by the Governing Board of the National Research Council, whose members are drawn from the councils of the National Academy of Sciences, the National Academy of Engineering, and the Institute of Medicine. The members of the committee responsible for the report were chosen for their special competences and with regard for appropriate balance.

This study was supported by Contract No. SES-0630359 between the National Academy of Sciences and the National Science Foundation; by Contract No. 7275 with the William T. Grant Foundation; by Contract No. 2006-7875 with the William and Flora Hewlett Foundation; and by Contract No. 20070001 with the Spencer Foundation. Any opinions, findings, conclusions, or recommendations expressed in this publication are those of the author(s) and do not necessarily reflect the views of the organizations or agencies that provided support for the project.

International Standard Book Number-13: 978-0-309-26161-6
International Standard Book Number-10: 0-309-26161-9

Additional copies of this report are available from the National Academies Press, 500 Fifth Street, NW, Keck 360, Washington, DC 20001; (800) 624-6242 or (202) 334-3313; http://www.nap.edu.

Printed in the United States of America

Suggested citation: National Research Council. (2012). *Using Science as Evidence in Public Policy.* Committee on the Use of Social Science Knowledge in Public Policy, K. Prewitt, T.A. Schwandt, and M.L. Straf, Editors. Division of Behavioral and Social Sciences and Education. Washington, DC: The National Academies Press.

THE NATIONAL ACADEMIES
Advisers to the Nation on Science, Engineering, and Medicine

The **National Academy of Sciences** is a private, nonprofit, self-perpetuating society of distinguished scholars engaged in scientific and engineering research, dedicated to the furtherance of science and technology and to their use for the general welfare. Upon the authority of the charter granted to it by the Congress in 1863, the Academy has a mandate that requires it to advise the federal government on scientific and technical matters. Dr. Ralph J. Cicerone is president of the National Academy of Sciences.

The **National Academy of Engineering** was established in 1964, under the charter of the National Academy of Sciences, as a parallel organization of outstanding engineers. It is autonomous in its administration and in the selection of its members, sharing with the National Academy of Sciences the responsibility for advising the federal government. The National Academy of Engineering also sponsors engineering programs aimed at meeting national needs, encourages education and research, and recognizes the superior achievements of engineers. Dr. Charles M. Vest is president of the National Academy of Engineering.

The **Institute of Medicine** was established in 1970 by the National Academy of Sciences to secure the services of eminent members of appropriate professions in the examination of policy matters pertaining to the health of the public. The Institute acts under the responsibility given to the National Academy of Sciences by its congressional charter to be an adviser to the federal government and, upon its own initiative, to identify issues of medical care, research, and education. Dr. Harvey V. Fineberg is president of the Institute of Medicine.

The **National Research Council** was organized by the National Academy of Sciences in 1916 to associate the broad community of science and technology with the Academy's purposes of furthering knowledge and advising the federal government. Functioning in accordance with general policies determined by the Academy, the Council has become the principal operating agency of both the National Academy of Sciences and the National Academy of Engineering in providing services to the government, the public, and the scientific and engineering communities. The Council is administered jointly by both Academies and the Institute of Medicine. Dr. Ralph J. Cicerone and Dr. Charles M. Vest are chair and vice chair, respectively, of the National Research Council.

www.national-academies.org

Preface

The Division of Behavioral and Social Sciences and Education (DBASSE) of the National Research Council (NRC) in 2005 established a standing committee to consider questions of how to strengthen the quality and use of social science research and to lay a foundation for the continuous improvement in the conduct of social science research and its applications to public policy. The standing committee was to identify areas of significant interest to those in the policy, research, and practitioner communities.

That committee convened a number of workshops and discussion meetings and met with a variety of researchers engaged in research on evidence for public policy, and it also consulted with policy makers about the usefulness of social science research to their work. As a result of those workshops and meetings, the committee concluded that it should give less attention to how the social sciences produce knowledge about policy, and focus, instead, on the settings and conditions that affect whether social science knowledge is used in policy making. To carry out the task identified, the NRC in 2009 set up the Committee on the Use of Social Science Knowledge in Public Policy.

This new committee decided to propose a framework for research on how policy makers make use of scientific knowledge and how the results of that research might lead to improved policy making and improved preparation of students in policy schools for careers in the policy world. This report is the result of the work of the committee.

In acknowledging the many people who made our work possible, we begin with thanks to members of the original standing committee (who are not members of the present committee), whose insights contributed to the work that followed: Thomas D. Cook, Northwestern University; Judith Feder, Georgetown University; Elinor Ostrom, Indiana University; Michael Peckham, University College, London; and Philip E. Tetlock, University of Pennsylvania.

We also acknowledge with gratitude and sadness the contributions of Stephen H. Schneider, Stanford University, who served on the standing committee and on this committee until his death in 2010.

Many NRC staff helped to guide the work of the standing and authoring committees. We particularly acknowledge the leadership of Michael J. Feuer, the former executive director of DBASSE, who guided the initial development of our study. We also thank other staff who helped the standing committee: Marty Orland and Catherine Freeman, initial study directors; Tina Winters, senior program associate; and Dorothy Majewski, administrative assistant.

As the co-editors drafted text for the report, one committee member, Norman Bradburn, met frequently with us, offering invaluable advice. Mary Ann Kasper served as our senior project assistant, and Viola Horek provided important administrative support. Editorial assistance was provided by Eugenia Grohman, and Kirsten Sampson-Snyder marshaled our report through review.

We also acknowledge the sponsors of this study: the U.S. National Science Foundation, the William T. Grant Foundation, the William and Flora Hewlett Foundation, and the Spencer Foundation.

This report has been reviewed in draft form by individuals chosen for their diverse perspectives and technical expertise, in accordance with procedures approved by the NRC's Report Review Committee. The purpose of this independent review is to provide candid and critical comments that will assist the institution in making its published report as sound as possible and to ensure that the report meets institutional standards for objectivity, evidence, and responsiveness to the study charge. The review comments and draft manuscript remain confidential to protect the integrity of the deliberative process.

We thank the following individuals for their review of this report: Robert F. Boruch, Graduate School of Education and Statistics, University of Pennsylvania; David S. Cordray, Center for Evaluation Research and Methodology, Institute for Public Policy Studies, Vanderbilt University;

James E. Geringer, Policy and Public Sector Strategies, ESRI, Inc., Cheyenne, Wyoming; Arthur Lupia, Institute for Social Research, University of Michigan; Robert A. Moffitt, Department of Economics, Johns Hopkins University; William D. Nordhaus, Department of Economics, Yale University; Michael J. O'Grady, National Opinion Research Center, Bethesda, Maryland; Lant Pritchett, Harvard Kennedy School, Harvard University; Daniel R. Sarewitz, Center for Science, Policy, and Outcomes, Columbia University; Bernard Silverman, Department of Statistics, Oxford University; and Laura Siminoff, Department of Social and Behavioral Health, Virginia Commonwealth University.

Although the reviewers listed above have provided many constructive comments and suggestions, they were not asked to endorse the conclusions or recommendations nor did they see the final draft of the report before its release. The review of this report was overseen by Lawrence D. Brown, Department of Statistics, The Wharton School, University of Pennsylvania, and Richard J. Bonnie, Institute of Law, Psychiatry, and Public Policy, University of Virginia. Appointed by the National Research Council, they were responsible for making certain that an independent examination of this report was carried out in accordance with institutional procedures and that all review comments were carefully considered. Responsibility for the final content of this report rests entirely with the authoring committee and the institution.

Kenneth Prewitt, *Chair*
Miron L. Straf, *Study Director*
Committee on the Use of Social Science
Knowledge in Public Policy

Contents

Summary

Fresh from the contributions made by science to the World War II success, at mid-century the nation adopted a broad policy to invest heavily in science and technology as a foundation for economic growth, social welfare, and national security. The emphasis was on the physical and biological sciences, but the social sciences were mobilized with respect to selected foreign and domestic challenges—area studies for the former and large-scale empirical projects on social welfare for the latter.

The 1966 study *Equality of Educational Opportunity* (known as the Coleman report) is a convenient marker for the arrival of "big" social science. It was designed to inform national and state policy relevant to reducing racial disparities in public education. Other large-scale research projects followed: on a negative income tax, housing allowances, and health insurance, among others. Evaluation research was announced as a new research specialty. Later in the century emphasis was placed on performance metrics, social indicators, ranking schemes, comparative assessment, and related tools and concepts based in social science. Private-sector organizations—university centers and institutes, think tanks, survey houses, and for-profit consulting firms—rapidly expanded in numbers and scope, as did graduate-level schools to prepare professionals for careers in public policy. The federal statistical system made available its significant information base for policy analysis in these nongovernmental settings. The federal government recruited social scientists in executive agencies and on congressional staffs.

By the end of the 20th century, a multibillion dollar policy enterprise was in place, drawing on private philanthropic support as well as federal and state funding. The task of this loosely interconnected policy enterprise is to describe social conditions, advise on policy interventions, test alternative program designs, and evaluate outcomes. This work is funded on the promise that good science will be used to decide what social conditions need attention, what should be a public responsibility or better left to the market or not-for-profit actors, and what interventions—to grow the economy, improve welfare, protect security—are efficient and effective.

As the policy enterprise expanded and extended its reach, interest mounted in whether its knowledge products were being adequately used. Continued investment in producing the knowledge suggested it was used and valued, but how valuable and for whom was uncertain. This uncertainty was addressed in a 1978 National Research Council report, *Knowledge and Policy: The Uncertain Connection.* The report found that, despite numerous social science studies of policy interventions and steps to increase their relevance to and use for policy making, "we lack systematic evidence as to whether these steps are having the results their sponsors hope for. . . ."

More than three decades later our report returns to the "uncertain connection," to again ask what is known about how scientific knowledge is used in public policy and how it can be more effectively used. The Committee on the Use of Social Science Knowledge in Public Policy was charged by the National Research Council "to review the knowledge utilization and other relevant literature to assess what is known about how social science knowledge is used in policy making . . . [and] to develop a framework for further research that can improve the use of social science knowledge in policy making."

The first charge, to assess what is known, led us to an early and obvious point. Knowledge from all the sciences is relevant to policy choices: the physical sciences inform energy policy on renewable efficiencies; the biological sciences inform public health policy on infectious diseases; the engineering sciences inform national defense policy on weapon design; the social sciences inform economic policy on international trade trends. Understanding whether, why, and how this scientific knowledge is used, however, is uniquely suited to the methods and theories of the social sciences. Making "use" of scientific knowledge is what people and organizations do. And what people and organizations do is the focus of social science.

To date, there has not been much success in explaining the use of science in public policy. We base this statement on three findings. First, although there are heuristically valuable typologies of ways science is used in policy, the typologies have not (and perhaps cannot) guide empirical research programs. Second, the research specialty labeled "knowledge utilization" has focused on challenges highlighted by the "two communities" metaphor (researchers and policy makers, each with their distinctive cultures) and proposed various innovations to improve communication and interaction between science and policy—brokering, translation, interaction models. There is little systematic research on whether these innovations are improving the use of science in policy, although there are clear indications that they are being usefully applied in practice settings. In fact, it is not even clear that the two communities metaphor is the most fruitful way to frame a study of knowledge use in policy. Third, although the relatively recent approach known as evidence-based policy and practice, focused on improving understanding of "what works," has influenced the *production* of scientific knowledge, it has made little contribution to understanding the *use* of that knowledge. In some of its more prominent formulations the issue of "use," because it involves political and value considerations, is said to be outside the scope of evidence-based policy.

If more than three decades of worrying about science use in public policy has not produced satisfactory explanations, it may be that we have been looking in the wrong place—for a coherent typology of use or ways to bridge the gap between two communities. The committee turned its attention to a research framework that draws on recent developments in social science perhaps better suited to explaining the use of science in public policy.

The first step in constructing the framework reprised a familiar point. Science, when it has something to offer, should be at the policy table. But it shares that table with an array of nonscientific reasons for making a policy choice: personal and political beliefs and values are present, as are lessons from experience, trial and error learning, and reasoning by analogy. Obviously, political matters and pressures weigh heavily when policy choices are made. Nevertheless science is a unique voice. What science has to say about policy choices results from investigations governed by systematic and rule-governed efforts that guard against self-deception—against believing something is true because one wants it to be. Because science is designed to be disinterested, if a policy question involves what are the "real" conditions or what will "probably" happen if one policy is implemented instead of another, science is generally a more dependable and defensible guide than informed hunches, analogies, or personal experience. Also, at least in a democracy, political leaders are obliged to give reasons for their policy choices—the theory of democratic accountability underpins this obligation. These reasons often require science-based description of conditions needing attention and explanations of what is likely to happen (or did happen) because of a policy intervention.

Science has five tasks related to policy: (1) identify problems, such as endangered species, obesity, unemployment, and vulnerability to natural disasters or terrorist acts; (2) measure their magnitude and seriousness; (3) review alternative policy interventions; (4) systematically assess the likely consequences of particular policy actions—intended and unintended, desired and unwanted; and (5) evaluate what, in fact, results from policy.

Across all of these tasks, there are political and value considerations that are outside the scope of science. We acknowledge that and build it into the recommended research framework.

A FRAMEWORK FOR RESEARCH ON USE

Policy is made in many settings. It evolves from a many faceted social process involving multiple actors engaged in assembling, interpreting, and debating what evidence is relevant to the policy choice at hand, and then, perhaps, using that evidence to claim that a particular policy choice is better than its alternatives. This process is best understood as a form of policy argument or practical reasoning that is persuasive with respect to the benefit or harm of policy actions. Policy argument includes generalizations, extrapolations, assumptions, analogies, metaphors, anecdotes, and other elements of reasoning that differ from and can contradict scientific reasons. From this perspective, scientific knowledge is "evidence" when that knowledge is used in support of statements relevant to policy claims. "Evidence" does not reside only in the world where science is produced; it emerges in the political world of policy making, where it is interpreted, made sense of and is used, perhaps persuasively, in policy arguments. Evidence-influenced politics is suggested as a more informative metaphor, descriptively and prescriptively, than evidence-based policy.

Our research framework argues for more careful study of policy argumentation, as well as for increased roles for the psychology of decision making and for systems perspectives. The social sciences offer important knowledge about how mental models, belief systems, organizational rules, societal norms, and other factors influence the behavior of decision makers. They also offer important knowledge about how people learn, when they optimize and when they satisfice; why they organize themselves, form institutions, communicate, establish norms, and develop routines; how they assess risks; and how they make decisions, individually and collectively. This array of scientific specialties has never fully addressed a key issue: when, why, how, even whether science is used in public policy making. Research can explain the cognitive operations and biases that policy makers and scientists bring to their work and the context-specific situations, practices, logics (ways of reasoning and understanding), and cultural assumptions of the settings in which they operate. Relevant research fields include social psychology, behavioral economics, decision theory, and organizational sociology. We urge scholars in these and related specialties to investigate the use of scientific knowledge in policy making.

Policy interventions unfold in large, complex, dynamic social systems. A systems perspective helps decision makers and researchers think broadly about the many effects a policy may produce and the ways in which a planned social intervention interacts with other existing interventions and institutional practices. Rarely can the study of the individual components of a system lead to a full understanding of the system. There are systems effects on individual actors and the system as a whole, including emergent, indirect, and delayed effects, as well as unintended and unpredictable consequences from the interactivity of a system's elements. The social sciences bring a variety of approaches and methodologies to the study of complex systems. Examples of the use of systems thinking in the study of national security, obesity prevention, and the evaluation of complex social interventions illustrate its potential utility in policy making more broadly.

THE NEXT GENERATION OF RESEARCHERS AND PRACTITIONERS

The three actors central to advancing and applying the research framework are established scholars in the fields and specialties identified above, Ph.D. candidates in those fields and specialties, and administrators and faculty responsible for curricula in schools and programs characterized by the term "policy education."

Established scholars have long-range research agendas and are not easily persuaded to drop them to pursue new questions. New research fields nevertheless emerge when even a few established scholars focus their theories and methods on a major question getting little attention. Among decision-making theorists, cognitive psychologists, and scholars of system properties are some, we expect, who will find that posing the issue of science use as needing *their* attention will be attractive. It is exciting to be in on the ground floor of a new field of research, especially when there is a large and influential audience waiting for guidance on how to strengthen science use in policy.

A companion effort focuses on students at the Ph.D. stage. There are numerous examples of philanthropic and federal funding that shaped the choice of dissertation topics and the early research trajectory of young scholars—resulting in new scholarly fields of inquiry. With heightened political attention to the "broader impacts" of science, answers are being sought, for example, for better ways to link natural and social sciences in addressing policy challenges, to better understand how variability in the quality of scientific evidence affects its use, and to the value of investing in intermediaries promising to promote the use of science as evidence.

The third audience is those responsible for the curriculum in public policy schools and programs in U.S. universities, which annually graduate thousands of students, many of whom find positions in the policy enterprise. It would be useful to know in some detail, first, what these students are and are not being taught as it bears on the use of science in public policy. That would require an investigation far more extensive than the committee could undertake. We did conduct a limited review sufficient to reach conclusions relevant to what our findings imply for policy education.

Our point is simple: policy education should equip its graduates to promote the use of science in policy-making settings. Graduates need, obviously, a working familiarity with the substance of policy issues and competency to locate, assess, and introduce validated research on those issues. But more is needed. Success at promoting science depends on grasping the complexity of the policy world, and on understanding the assumptions underlying divergent policy framings, expert judgments, and consensus-building techniques, as well as standard analytic methods and approaches. Policy students can be taught to appreciate policy making through policy argument or practical reasoning and to understand that the relevance of and weight given to science depends on the policy context. They can recognize the limits of the persuasive power of scientific reasoning, the substantial institutional barriers and cultural resistance to new scientific knowledge, and the role of moral and ethical beliefs.

For a century or more the social sciences have contributed to policy making in many ways, especially in informing policy design and evaluation. We see a fresh way they can further contribute: by specifically focusing on whether, why, and how science is, or is not, used as evidence in public policy.

1

Introduction

T his report is about using science as evidence in public policy. Science identifies problems—endangered species, obesity, unemployment, and vulnerability to natural disasters or bioterrorism or cyber attacks or bullying. It measures their magnitude and seriousness. Science offers solutions to problems, in some instances extending to policy design and implementation, from improved weapons systems to public health to school reform. Science also predicts the likely outcomes of particular policy actions and then evaluates those outcomes, intended and unintended, wanted and unwanted. In these multiple ways science is of value to policy, *if used*.

FOCUS OF THE REPORT

The report title—"using science as evidence in public policy"—takes on a specific meaning in this report. Policy makers offer reasons for their policy actions, reasons that bear on whether to take action at all, that address the interests and values at stake, and that claim the policy will work as intended, without unwanted consequences. These reasons are embedded in a policy argument; and a policy argument, to borrow a term from philosophy, is a form of practical reasoning. The term "argument" here has no pejorative implications. A policy argument is intended to persuade others to accept the reasons supporting or opposing a policy action.

In this report, a general term, "using science in public policy," has a precise meaning: knowledge based in science is presented as evidence to support reasons used in a policy argument. Knowledge based in science is broadly taken to mean data, information, concepts, research findings, and theories that are generally accepted by the relevant scientific discipline. Science is not the only source of knowledge used in policy argument—beliefs, experience, trial and error, reasoning by analogy, and personal or political values are also used in policy argument. How science interacts with nonscientific reasons given for public policies is among the issues we address, especially the complicated but inevitable interaction of politics, values, and science.

"Use" is another key term in the report. We review how it is defined and studied in the research specialty known as knowledge utilization. We consider what is known about if, when, and why use occurs, the various efforts to improve use, and how the current interest in evidence-based policy relates to use. The report focuses on what is poorly understood about use and might be better understood if social science research shifted its focus from defining use to studying what occurs in policy arguments when relevant science is available.

"Policy" is broadly construed in this report. It is used to describe specific and detailed adjustments to established policies, such as modifying the rate at which capital gains are taxed. It is also used for more general topics, such as school reform or deficit reduction, each of which can encompass dozens of discrete policy choices and instruments. And it is used even more broadly to reference policy domains, such as welfare policy or security policy. We even stretch the term to include the broadest of national policy goals, such as strengthening the market economy or protecting the civil rights of all Americans, which involve hundreds of discrete policies adopted and modified over decades. The general principles laid out in this report would be applied differently depending on the level of policy specified, on the particular policy sector (e.g., social welfare or national security) and on whether the policy target is a current condition, such as stopping illegal immigration, or one anticipated years or even decades hence, such as future energy needs in a world of 9 billion people. These differences matter, but we do not take them up. We consider what it means for science "to be of use" in a framework that does not depend on a carefully formulated definition of policy.[1]

---

[1]We restrict attention to the use of science in government public policy. There are of course other arenas where policies with public consequence are made—business policies about product lines or investment strategies, university policies about diversity initiatives or tenure criteria, and advocacy group policies about pressure tactics or fundraising goals. Although points made in this report are applicable beyond the arena of government policy, this is not our topic.

It is also important to say what the report is not about. It is not about the impact of science on society or about the payoff of investing in social science. These issues are being actively discussed in leading scientific institutions and in funding agencies, and we discuss this heightened interest in Chapter 2. Clearly, unused science cannot have any impact, but use does not equal impact. To assess impact and, beyond impact, return on investment, requires analysis beyond the scope of this committee's charge. Our focus is restricted to use.

AUDIENCE

This report is addressed to scientists in general and to social scientists in particular. The use of science as evidence in policy making—irrespective of its disciplinary source—is a social phenomenon, and therefore a proper object of analysis for the social sciences. We present a research framework that can improve the scientific understanding of the use of science in public policy. Although some argue that the improved use of science will lead to improved policy choices, that is not our claim here. The question of what "improved" policy or "better" policy making entails and on what criteria such improvements might be judged is beyond our scope. What science does, with lesser to greater certainty and confidence, is describe conditions of interest to policy makers (or that might come to interest them when they are described), probe into natural and social conditions that may give rise to the need for policy action, predict what is likely to happen if action is taken (or not taken) to address those conditions, and, once an action is taken, explain what did happen and why.

Scientists—when they are practicing science—do not tell policy makers what should interest them or what policy choices they should make. Scientists deal with accurate description of conditions and with explanations about the causes or consequences of those conditions. Physicists and mathematicians at Los Alamos estimated the destructive consequences of the atom bomb. Social scientists in the Office of Strategic Services (predecessor to the Central Intelligence Agency) estimated the bomb's effect on Japan's civilian morale. Scientists could say, with varying degrees of certainty, that, *if* an atomic bomb is dropped, the consequences are likely to be this rather than that. There was no scientific basis on which to say *whether* to drop the bomb. That decision fell to President Harry S. Truman and his political and military advisers, who had to weigh factors in addition to those based in science.

Science does, however, bring one special asset to the table. It is a process of producing knowledge directed by systematic and rule-governed efforts that guard against self-deception—against believing something is true because one wants it to be true. We are not claiming that scientists are immune to self-deception; we are claiming that correctly doing science results in disinterested knowledge. For this reason, when the question on the table is what are the "real" conditions or what will "probably" happen if we implement one policy instead of another, science is on balance a more dependable and defensible guide than informed hunches, analogies, or personal experience.

Dependable and defensible does not equal certainty. Science is always uncertain and can, over time, be wrong—19th century race science, for example. But, of course, no source of knowledge or mode of reasoning escapes uncertainty and error when it comes to assessing what policies do or fail to do. Scientific investigations—whether in geology, biochemistry, epidemiology, or sociology, and across the policy issues each addresses, from toxic waste disposal, to bioterrorism, infectious diseases, and social violence—will, on balance, be a more dependable ground on which to argue that a policy action will or will not have certain effects than other sources of knowledge. Whether policy makers use the results of scientific investigations is an altogether different matter, and the subject of this report.

UNDERSTANDING THE SOCIAL SCIENCES AND THEIR ROLE

There are several social sciences and an even greater number of methods, approaches, theories, and research strategies in something as broad and indeterminate as understanding the human condition. What the social sciences share is their analytic focus on the behavior, attitudes, beliefs, and practices of people and their organizations, communities, and institutions. The social sciences study social phenomena, including social phenomena conditioned and caused by or responsive to matters that are investigated in the natural sciences—earthquakes, infectious diseases, ocean currents.

In associating the label *science* to specialties ranging from cultural anthropology to neuropsychology, we use the term differently than it is used by disciplines, such as physics or chemistry, which have a well-developed set of comprehensive, generative theories that both explain and predict phenomena. Social science may be understood by some of its practitioners in this way, but we favor what is indicated by the German term "*Wissenschaft*" and its linguistic equivalents that refer to any disciplined, systematic inquiry with established methods and rules of evidence and inference that protect the investigator from self-deception.[2]

Many conditions at stake in a policy choice are not social—collapsing bridges, atmospheric pollution, species loss. Evidence from engineering, chemistry, and ecology describes those conditions and their causes. Yet even when the policy is about physical or biological conditions, the need to consider the human actor is seldom absent when considering policy options. Biochemistry and epidemiology show that smoking is dangerous to health; different social sciences assess policy options to reduce tobacco use: increasing the cigarette tax (economics), restricting where people can smoke (political science, social psychology), requiring warning messages (social psychology). Geology and physics assess the leakage risks of storing nuclear waste at a proposed repository, but safety also depends on a warning symbol that can communicate radiation danger for thousands of years, and, for that, linguistics, anthropology, and other social sciences involved in risk communication are needed. There is a large and growing list of policies guided by natural and social science. Topics in the disciplines of science, technology, engineering, and mathematics are matters for experts in these fields. What topics can be taught, at what levels, and how to teach the topics effectively are matters for educational psychologists and learning experts.

We begin to see that there are two ways in which social science matters to policy. First, social science contributes to understanding conditions and consequences of concern to policy makers; second, social science has methods and theories applicable to investigating the use of science in policy. Use, we have said, is itself is a social phenomenon. Use occurs in specific kinds of social organizations—executive agencies, legislatures, or expert committees—each conditioned by organizational norms, cultures, and patterns of interaction that are studied in sociology, social psychology, and organizational specialties. Use involves political choices in a wide variety of policy settings and thus is a topic for researchers in political science and public administration who investigate policy networks, intermediaries, lobbyists, knowledge brokers, and institutional rule making. Use is a particular kind of decision making and is examined with concepts from philosophy, such as argumentation and practical reasoning, as well as psychological theories, such as behavioral decision theory. Use depends on users learning

---

[2]Science fraud is a deliberate effort to deceive others, to persuade them to believe what is known to be false. Fraud is not our concern in this report, except to make the obvious point that it can undermine the confidence of policy makers looking at scientific evidence and not knowing if it is responsibly or fraudulently produced.

what sources of knowledge are dependable guides, and is investigated using cognitive theory at the individual level and sociocognitive theory at organizational levels. Use is highly contextual, conditioned by situated norms and habits, and is studied anthropologically and sociologically. Finally, use of science in policy can be seen as selecting among bodies of knowledge or expert opinion; it is then a topic in the sociology of knowledge, including science and technology studies.

In summary, the social sciences have two responsibilities. The first is to accurately describe human behavior and social conditions, including their causes and consequences, and, when policies are implemented to change those behaviors and conditions, to assess the consequences. This responsibility is most frequently discussed as social science investigation of behavior and social conditions. But we emphasize that the responsibility extends to many policies that address natural conditions, when the policy intends, anticipates, or will be affected by changes in human behavior and social structures.

The second responsibility of the social sciences is to focus their formidable array of methods and theories on understanding how social *and* natural scientific knowledge is used as evidence in the policy process. This responsibility is anticipated in the committee's statement of task (see Box 1-1) and developed in detail in the report.

BOX 1-1
Statement of Task

The committee will develop a framework for further research that can improve the use of social science knowledge in policy making. The committee will review the knowledge utilization and other relevant literature to assess what is known about how social science knowledge is used in policy making. The framework will indicate the potential for new ways of understanding the use of social science knowledge in policy making. The framework will also have implications for the content and scope of training in schools and programs that prepare students for careers that use social science knowledge in policy making.

THE ROLE OF POLITICS AND VALUES
IN UNDERSTANDING USE

A familiar argument views science as a means of rescuing policy from short-sighted influence peddling and power politics (DeLeon, 1988; Dryzek and Bobrow, 1987; Majone, 1989; Stone, 2001). The view that science can be a counterweight to self-interestedness in politics and thereby ensure that policy reflects the public interest has a distinguished tradition, dating to the American progressive movement and famously voiced even earlier by Woodrow Wilson (1901) in his Ph.D. thesis, *Congressional Government: A Study in American Politics*. That view—which could be found as well in the early 20th century among English new liberals and European Christian and social democrats—held that modern knowledge of society, grounded in the new social sciences, could generate useful policy ideas based on putatively objective and factual bases. Henig (in press) has described the influence of this way of thinking on education policy:

> The argument that politics is the enemy to be kept at bay has been influential in shaping America's thinking and its actions, both historically and on the contemporary scene. It informed and justi-fied structural changes successfully promoted by the Progressive Reformers of the early 20th century. "There is no Democratic or Republican way to pave a street," was a slogan of the time, with the implication that there was, instead, an objectively correct way, best determined via technical and scientific expertise. Policies like teacher certification, civil service protections, and the formal assignment of education policy making to school boards indepen-dent from municipal governments and the political machines that often controlled them were portrayed as a way to empower the experts, who would both know and respect objective data, and ex-plicitly buffer them from political interference, patronage politics, and faddish and emotion-driven popular whims.

This tradition has contemporary adherents. The Urban Institute, in making the case for evidence-based policy, states that a "question that figures into all public policy decisions—What political and social values do the proposed options reflect?—is largely outside the scope of evidence-based policy" (Dunworth et al., 2008, p. 1). The hope that science could be

separated from politics is summarized (although not endorsed) by Deborah
Stone (2001, p. 376):

> Inspired by a vague sense that reason is clean and politics is dirty,
> Americans yearn to replace politics with rational decision-making.
> Contemporary writings about politics, even those by political
> scientists, characterize it as "chaotic," "the ultimate maze," or "or-
> ganized anarchy." Politics is "messy," "unpredictable," an "obstacle
> course" for policy and a "hostile environment" for policy analysis.
> . . . Policy is potentially a sphere of rational analysis, objectivity,
> allegiance to truth, and the pursuit of the well being of society as a
> whole. Politics is the sphere of emotion and passion, irrationality,
> self-interest, shortsightedness, and raw power.

Holding to a sharp, a priori distinction between science and politics is
nonsense if the goal is to develop an understanding of the use of science in
public policy. Policy making, far from being a sphere in which science can
be neatly separated from politics, is a sphere in which they necessarily come
together (Jasanoff, 1990). As suggested in the Urban Institute quotation,
"evidence-based policy" stops where politics and values start. Our position
is that the use of that evidence or adoption of that policy cannot be studied
without also considering politics and values.

For both descriptive and prescriptive reasons, then, *evidence-influenced
politics* is a more informative formulation than evidence-based policy. It
is descriptively informative in the sense that it occurs whenever scientific
evidence enters into political deliberations about policy options, and this
occurs much more regularly than the apolitical, narrowly focused activities
characteristic of evidence-based policy. We support this assertion throughout this report, starting below in the section on democratic theory. Evidence-influenced politics is also prescriptively important. Policy routinely
involves value and related considerations that are outside the expertise of
science. Even when values are at stake, scientists can legitimately advocate
for attending to knowledge that accurately describes the problem being addressed or that predicts probable consequences of proposed actions. It is our
normative position that if policy makers take note of relevant science, they
increase the chances of realizing the intended consequences of the policies
they advance. This is evidence-influenced politics at work.

The relative weight in any policy choice of the three strong forces—

political considerations, value preferences, scientific knowledge—shifts depending on many factors; a short list includes

- the accuracy and persuasiveness of the descriptive analysis of the targeted social condition;
- the reliability of instruments and data sets used to assess the magnitude, gravity, and trajectory of the condition;
- the level of certainty about the direction and strength of causal inferences linking intervention to desired outcome;
- whether the task is evaluating what has happened or is estimating what will happen;
- the weight accorded to knowledge that comes from experience and practical expertise;
- the level of concerns about unwanted or unplanned consequences;
- the social values at stake, and how widely they are shared; and
- the power base of organized political interests.

Some mixture of politics, values, and science will be present in any but the most trivial of policy choices. It follows that use of science as evidence can never be a purely "scientific" matter; and, it follows that investigating use cannot exclusively focus on the methods and organizational settings of knowledge production or on whether research findings are clearly communicated and how.

POLICY MAKING IN A REPRESENTATIVE DEMOCRACY

Rigorous investigation of how science is used in the United States has to start with the principles and realities of the nation's democratic politics. Obviously our treatment of such a vast terrain is highly selective, commenting on only a few issues to illustrate a broader point: there is no way to examine "using science in public policy" apolitically. Our selective entry point is the theory of democratic accountability. This theory emphasizes electoral competition among ambitious people who want power and want to retain it after they get it. (See Schumpeter, 1942, for a representative treatment of this theory.) To realize their political ambitions, aspiring or incumbent leaders "count the votes." This is critical to democratic accountability. When leaders are indifferent to the strength of their political support, the link between democratic accountability and elections is correspondingly weaker. Making policy choices based, even in part, on gaining or retaining majority support is, for Schumpeter and others, a necessary feature of democratic accountability. Counting the votes, however, can lead to "ignoring the evidence" about policy consequences in favor of responding to voter preferences. The tension in choosing between being a trustee of the public good or a delegate responsive to one's voting constituency—eloquently expressed by Edmund Burke in the 19th century—is inescapable in a democracy.

A similar logic holds for interest group politics. Politics enters the policy process through organized interests, which invest resources—estimated at $3.49 billion in 2010 (Center for Responsive Politics, 2011)—to directly influence policy.[3] This process, like electoral politics, may ignore, downplay, distort, or vociferously contest scientific knowledge that fails to support a group's desired policies. But the suppression of interest groups' preferences is not an option in a functioning democracy. Institutional arrangements in democracies are, after all, designed around the assumption that policy choices are contested.

Democratic political theory also places values at the center of politics. Esterling (2004) contrasts normative and instrumental reasoning, making the point that arguments for why a policy is desirable or undesirable can be made independently of its immediate social consequences. Legislators might agree with science showing how mandating helmets for motorcyclists reduces highway fatalities, and yet disagree about whether to "use" the science. To accuse a libertarian who prefers minimal government and maximum individual choice of "ignoring the evidence" about fatality rates misses the point. Just as electoral calculations and interest considerations cannot be suppressed in a democracy, neither can value preferences. In fact, political principles, such as the first amendment, are designed to promote forceful value expression.

The neoconservative critique of the social welfare state blended scientific and normative arguments. Wilson (1996, p. viii) described the law of unintended consequences as an "article of faith common to almost every adherent" of neoconservatism:

> Things never work out quite as you hope; in particular, government programs often do not achieve their objectives or do achieve them with high or unexpected costs. . . . Neoconservatives, accordingly, place a lot of stock in applied social science research, especially the sort that evaluates old programs and tests new ones.

---

[3]See Center for Responsive Politics. *Lobbying Database.* Available: http://www.open secrets.org/lobby/ [August 2012].

Other voices in the neoconservative movement, with a less scientific bent than Wilson, simply started from the premise that the market is superior to the state in producing solutions to social problems ranging from poverty to education. The Heritage Foundation writes that its mission is "to formulate and promote conservative public policies based on the principles of free enterprise, limited government, individual freedom, traditional American values, and a strong national defense."[4]

If democratic politics invites competition for power, contesting interests, and the expression of diverse values—all of which interact in complicated and not always welcoming ways toward science at the policy table—another feature of democracy more clearly does open space for science. Democracy rests on the obligation of rulers to give reasons for policies. It is not acceptable to say "Fight this war or pay this tax because I am your ruler and I say so." The obligation to provide reasons generally involves explaining that a given policy will prevent a social harm or advance a desired public welfare goal—such as why one public health intervention rather than another saves lives, why security practices are needed to protect against terrorism, or why increasing teacher salaries will improve educational outcomes. When there is a scientific basis for a proposed policy—about the effectiveness of a vaccine or the deterrent effect of airport security or the correlation between teacher pay and student performance—and the reason given for the policy is the effects it will produce, the use of science provides more dependable as well as more defensible reasons than does unsupported presumption or speculation.

Here, however, we again emphasize that a dependable and defensible reason will not necessarily be used just because it is available. Re-election concerns, interest group pressure, and political or moral values may be given more weight and may draw on reasons outside the sphere of what science has to say about likely consequences. A democracy as readily allows the conservative mission of the Heritage Foundation noted above as it does the liberal agenda of the Center for American Progress, which is "dedicated to improving the lives of Americans through progressive ideas and action."[5]

We summarize this brief foray into democratic theory with a current policy debate: school choice. It was not inherent in this issue that it be framed as one putting "market solutions" on one side of an ideological divide and "government's responsibility for public welfare" on the other.

---

[4]See http://www.heritage.org/ [January 2012].
[5]See http://www.americanprogress.org/ [January 2012].

Charter schools, for example, were initially favored by educators and parents in order to escape "rigid and monotone bureaucracies, to be free to start schools employing innovative pedagogies, to allow families having a bad experience with their neighborhood school to look for a better fit for their child without having to exit the public system" (Henig, 2009, p. 148). Conservative foundations, which had been advocating for a universal school voucher system, turned to charter schools as a better test case for claiming that market choice was inherently superior to government provision of social services, including education. Advocates on the left, who might otherwise have defended charter schools as a progressive public-sector reform, opposed them in making "a tactical decision to fight the battle on this market *versus* pubic education ground" (Henig, 2009, p. 148). This tactical decision rested on the assumption that Americans had a deep allegiance to public education.

This was democratic politics at work. Partisan and ideological lines formed and hardened in ways that affected the role of science. Prospects "quickly faded that research could easily and simply unfold, methodologically and systematically driven by its own internal logic" (Henig, 2009, p. 148). Instead, research became enmeshed in the battle over clashing values and partisan interests.

Yet that is not the entire story. Researchers who sharply differ on whether charter schools yield positive effects, attacking each other's methods in the process, nevertheless agree on an important common and by now familiar finding. Factors outside the school, most particularly the role of family and community, account for more of the variation in school outcomes than do a school's characteristics, in this case, charter schools or traditional public schools.

> [T]he core of the research enterprise has not been corrupted . . . below the radar screen the collective enterprise of research is performing more or less as we might hope it would. . . . Good studies, as they accumulate, are pushing weaker studies to the margins, and studies claiming large, uniform, and unambiguous results are in some instances revealed to be unreliable outliers. (Henig, 2009, p. 143)

In the charter school example, all three forces—politics, values, science—are in the mix. The use of science cannot but be affected by how a policy issue is framed, and that initial step is largely beyond the reach of science. Yet science as it accumulates can reduce the range of political disagreement.

A BETTER GUIDE, NOT A BETTER POLICY

Commentary on the use of science in public policy frequently argues that its use will produce better policy or improve policy making. We offer a narrower but, we believe, more scientifically sound position, particularly with reference to the social sciences. Social science does not promise "better policy." It is not social engineering, misguided accusations notwithstanding. It is, simply, a guide to understanding problems, the conditions that give rise to those problems, and the outcomes likely to occur when policy addresses those problems. In this very specific sense, social, as well as natural sciences, are a more reliable ("better") guide than what is otherwise available to policy makers in considering many issues.

REPORT STRUCTURE

The United States has established a loose but large network of institutions and practices focused on providing scientifically grounded descriptions and causal explanations of conditions that are or could become the object of policy attention. The next chapter uses the shorthand term "policy enterprise" to describe this network. Its workings, its funding, and its purposes are the proximate context for a fresh examination of the science-policy nexus generally and the issue of use in particular.

Chapter 3 moves to the substantive material of the report, reviewing how knowledge use has been studied over the last half-century, what has been learned from that research effort, and what remains poorly understood. Chapter 4 presents a research framework, briefly summarizing selected concepts and research fields—especially related to practical reasoning, cognitive and social psychology, and systems thinking—for their application to deepening understanding of how science interacts with policy. The final chapter explains who needs to do what to advance the research framework outlined in Chapter 4. Appendix A reviews selected research methods that are particularly appropriate for research related to public policy when the social science task is to describe causes and consequences of social conditions and to assess the outcomes when policy tries to change those conditions. Appendix B contains the biographical sketches of committee members and staff.

2

Why This Report Now

Using science in public policy is on the nation's agenda. One reason is the growing demand for performance measures and enhanced accountability in federal agencies and not-for-profit organizations. Another is the call for evidence-based policy and practice, part of a broader focus on data-driven decision making across government agencies.[1] And in a period of fiscal restraint there is pressure to demonstrate that government-supported science offers benefit to taxpayers, a matter often discussed as "broader impacts."

"Broader impact," when associated with the social sciences, is typically understood as being useful for policy. There is an extensive network of intermediary organizations dedicated to bringing research knowledge to the world of policy making. A growing number of policy schools and programs are preparing students for careers in these many nongovernmental organizations, as well as government agencies and corporations, whose strategies depend on reliable knowledge of the society, economy, and polity. We summarize all of this as a policy enterprise, which is defined in more detail

---

[1] A memo from the U.S. Office of Management and Budget (May 18, 2012) to agency heads stresses that "Agencies should demonstrate the use of evidence throughout their Fiscal Year (FY) 2014 budget submissions. Budget submissions also should include a separate section on agencies' most innovative uses of evidence and evaluation . . . the Budget is more likely to fund requests that demonstrate a commitment to developing and using evidence." Available: http://www.whitehouse.gov/sites/default/files/omb/memoranda/2012/m-12-14.pdf [July 2012].

below. Actors in this policy enterprise have a stake in whether social science knowledge is useful to and used in policy making.

Although this report specifically addresses what is known and what needs to be studied about how social science is useful for policy making, the analysis bears on how evidence from all sciences is used. The stakes are high for every branch of science that claims to advance social welfare, contribute to economic growth, and enhance national security. The Introduction noted that "use," being a social phenomenon, is investigated with social science theories and methods, not with the theories and methods of physics, chemistry, biology, or engineering—even though these sciences place an enormous range of issues on the policy agenda. The social sciences should approach their responsibility to study "use" alert to consequences for the physical, biological, and engineering sciences as well.

"BIG" SOCIAL SCIENCE

Sustained attention to the use of social science in policy making received a noticeable boost in the post–World War II period, when leaders saw in "big science" a path to economic growth and social betterment. The U.S. National Science Foundation (NSF), created in 1950, is the premier institutional expression of this vision. The social sciences were initially excluded from the NSF, but a new role for social science nevertheless emerged (by the 1960s the NSF was funding social science). A convenient marker of the new role is the highly influential study of public schools known as the Coleman report (Coleman, 1966), undertaken in response to a 1964 congressional instruction that the commissioner of education investigate "the lack of availability of equal education opportunities for individuals by reason of race, color, religion, or national origin." By the standards of social science at the time, this study was big: 600,000 students and 60,000 teachers in 4,000 public schools.

Its size was only one of its distinctive characteristics. The study emphasized educational outcomes, breaking from a research tradition that had largely focused on inputs, such as expenditure per student. The Coleman study is best known for its controversial finding. Student test results and educational aspirations, which were the outcomes measured, could be explained as much by family background variables as by school characteristics (such as classroom size). As noted in the discussion of charter schools in the previous chapter, researchers are still actively investigating the subtleties of this key finding, as well as the report's companion finding that minorities enter school burdened with accumulated educational disadvantages.[2]

The Coleman study signaled that large-scale social science projects could inform the nation on critical policy challenges. A nation that used science to design radar and make the atomic bomb for its war effort could also declare a "war on poverty" and a "war on drugs," expecting the social sciences to help policy makers design programs and then evaluate whether programs were having their intended effect.

The government launched an ambitious agenda of nationally scaled social experiments based on randomized field trials. Examples included studies of a negative income tax, housing allowances, health insurance, and time-of-use electricity pricing. The government funded specialized institutes, such as the Institute for Research on Poverty at the University of Wisconsin. University-based survey capacities grew apace, notably the Institute for Social Research (ISR) at the University of Michigan and the National Opinion Research Center (NORC) at the University of Chicago. The federal government, the main provider of social and economic statistics, made its data easily available for analysis by university-based researchers, who in turn began to influence what statistical data were collected. A steady stream of studies based on secondary analysis of labor, health, income, crime, and related statistics from what has grown to nearly 90 federal programs and agencies underpinned debates about policy challenges and options. Hundreds of dissertations used federal statistics, and the writers of these dissertations became professors and researchers in university social science departments and interdisciplinary centers, where they produced the next generation of researchers trained to ask "big" questions about social welfare and economic trends, public health, and school reform. An early preoccupation in this research was whether the policies were having the expected outcomes.

Measuring outcomes quickly moved to the center of debate over the significant investments in the "Great Society" programs. The first *Handbook of Evaluation Research* was published in the mid-1970s (Guttentag and Struening, 1975). Policy and program evaluation focused attention on research at the intersection of what policy makers needed to know and what social science research offered. Is a program producing its intended

---

[2]See also Mosteller and Moynihan (1972), which arose from a Harvard faculty seminar to reassess Coleman's research. Although some conclusions were contrary to Coleman's findings, the reanalysis generally agreed with the relationship between educational achievement and equality of opportunity.

outcomes? Is it cost effective? Research in the 1970s challenged basic assumptions about the merits of the Great Society social programs, pointing out costly program failures and unintended negative consequences.

New ways of linking knowledge to policy making began to appear under formulations that are familiar today: evidence-based policy, performance metrics, impact assessment, and comparative effectiveness research. These formulations in turn led to institutional innovations, such as best practice guidelines and the philanthropically funded Coalition for Evidence-Based Policy.

The expanding influence of social science over the last half-century was aided by significant improvements in research methods. Advances in qualitative methods allowed researchers to examine complex problems with increasingly sophisticated case-study approaches, moving into research sites heretofore less carefully examined—including corporate decision making and laboratory science. There were significant advances in large-scale data collection, along with improved methods of analysis allowing policy analysts to handle census data and surveys with thousands of respondents and hundreds of variables. More recently, of course, there has been exponential growth in computing power and in analytic techniques. Further expanding the capacity of social science is the availability of administrative data and commercially collected digital data (Lazer et al., 2009). "Big data" is the recently coined term to describe, especially, the flood of data provided as a by-product of electronic media and transactions. Newly formed university centers and programs are actively exploring data visualization, data mining, and internet data in the new field of computational social science.

With huge amounts of accessible data, the technical knowledge to analyze the data, and hundreds of organizations seeking to link research to policy making, it is not surprising to find strong political interest in financial and performance audits, process monitoring, and impact evaluations, all of which are part of the broad interest in ways to hold officials and institutions accountable. Social indicators are pressed into service to describe what is going well, or not so well, in society, leading to such efforts as the "Measure of America" (Lewis and Burd-Sharps, 2010, Social Science Research Council, 2012), modeled on the United Nations' Human Development Index (United Nations, 2012), and the Key National Indicators Initiative (State of the USA, 2012). Ranking schemes, from corporate corruption to happiness to university performance, are ubiquitous in the media, their sometimes-questionable assumptions and methodologies notwithstanding.

This increasing sophistication of measurement and quantification does more than provide technical tools. It can have an independent effect on what kinds of policies reach the political agenda, and who is likely to be favored. When demographic analysis showed that racial minorities were undercounted at higher rates than the white population, the U.S. Census Bureau examined whether a statistical technique used in wildlife studies (dual-system estimation) could be used to adjust the count, to reduce, or even to eliminate this differential undercount. The research and methodological work led to an intense and highly partisan battle stretching across two decades, with litigation that reached the Supreme Court. It finally ended only when the Census Bureau leadership said that census adjustment was no longer being considered.[3]

School reform offers another example of policy options shaped by measurement decisions. The data-driven accountability movement in education challenges traditions—local control of schools, deference to professionalism, and the belief in public schools as a foundation for personal mobility and societal progress. Data-driven accountability "is not simply *affected* by this institutional upheaval . . . it is implicated in the upheaval itself" (Henig, in press). This occurs as accountability thinking and practice shifts power from local to state and federal levels, and undermines union control over teacher salaries and promotions.

Theories from political sociology explain these examples. In an influential essay, *Identity and Representation*, Bourdieu (1991, pp. 220-228) elaborates on how official measurement can constitute new policy realities. He develops his theory with reference to ethnic identity, described as "struggles over classifications, struggles over the monopoly of the power to make people see and believe, to get them to know and recognize, to impose the legitimate definition of the divisions of the social world and, thereby, to *make and unmake groups*" (p. 221 [italics in original]). Bourdieu adds that the power of statistical categories to move beyond being simply descriptive to being constitutive of social reality is proportional to the authority of agencies creating those categories—in his discussion, the government and the social sciences.

There are, then, many interlinked developments that establish the era

---

[3]The Supreme Court ruled that an adjusted count could not be used for apportionment, but it left open the question of whether an adjusted count could be used for other purposes, including redistricting and allocation of federal funds. Given the steadily increasing costs of the census and the persistence of errors—missed households and erroneous enumeration—it is likely that modifications or technical improvements on the basic design based on mailout/mailback and nonresponse household follow-up will continue to be a policy issue.

of big social science. Essentially, the country is replacing expensive trial-and-error policy making with more deliberately produced knowledge that can inform policy making (Ruttan, 1984, p. 552; cited in National Research Council, 1999b):

> Throughout history, improvements in institutional performance have occurred primarily through the slow accumulation of successful precedent, or as a by-product of expertise and experience. Institutional change was traditionally generated through the process of trial and error much in the same manner that technical change was generated prior to the invention of the research university, the agricultural experiment station, or the industrial research laboratory. With the institutionalization of research in the social sciences it is becoming increasingly possible to substitute social science knowledge and analytical skill for the more expensive process of learning by trial and error.

THE PRECURSOR TO BIG SOCIAL SCIENCE

In one important way, the last half-century of big science continued earlier understandings of what linked social science to policy. Even the immature disciplines of the early 20th century were mobilized to address national challenges in World War I, in fields as varied as propaganda analysis, competency testing, and economic planning. The monumental *Recent Social Trends*, commissioned by President Herbert Hoover in 1929 and financed by the Rockefeller Foundation, can be viewed as a precursor to big social science, though it did not have the impact on policy debate that later attended the explosion of large-scale research in the 1960s.

The national call on social science expertise was reinvigorated by the challenges of the depression economy in the 1930s, leading to major advances in micro- and macroeconomic analysis and in welfare policy initiatives, such as the Social Security system. Government's need for a better understanding of how the economy was responding to policy interventions led to improved scientific capacity, exemplified by population sample survey methods. World War II repeated the World War I call on social science expertise, but on a much larger scale. The Office of Strategic Services (which became the Central Intelligence Agency), for example, recruited historians, anthropologists, political scientists, and economists to help war planning in unfamiliar parts of the world, and psychologists to help break Japanese and German codes.

The period that spanned the two world wars established a basic pattern that set the stage for the policy enterprise. The government did not establish a significant internal capacity in the social sciences or even in policy analysis. In an early version of outsourcing, it turned to expertise outside of government. In this way, the first half of the 20th century established the principle that universities, specialized research organizations, and think tanks were a source of independent and nonpartisan social science knowledge on social conditions and policy options.

In these early decades, there was little attention to "use" as it is discussed today. Use was largely taken for granted, certainly in settings, such as the Social Science Research Council, the National Bureau of Economic Research, the Brookings Institution, and private foundations and funders. A widely discussed book, *Knowledge for What?* (Lynd, 1939), was subtitled *The Place of the Social Sciences in American Culture*. Today, of course, the subtitle would likely replace "in American culture" with "in policy making." The debate initiated by Lynd's book took place among social scientists and focused on the general public purposes for which research knowledge was generated, not on specific policy applications. A recurring theme in the debate was how to maintain scientific independence in order to "speak truth to power," a theme that has not entirely disappeared (O'Connor, 2007).

However, and indicative of a major shift in the thinking of U.S. social scientists since the 1930s, few scholars today want to keep their distance from the policy process or believe that independence can be secured only if their work lacks immediate relevance. When researchers are asked which they prefer, "more links between the academic and policy communities" or "a higher wall of separation," more than nine of ten opt for more links (Avey et al., 2012). This helps explain why the phrase "basic versus applied science," so pervasive in the immediate postwar period, does not appear in our report. Whether researchers call what they do basic or applied, they want it used—and for social scientists that means used in policy making. Policy makers looking for answers do not care whether social scientists call what they do basic or applied or, for that matter, care whether the research is disciplinary or interdisciplinary. These distinctions might still be of interest in some social science settings. They are not in this report, because we assume that policy makers are interested in answers to their questions, not in the particulars of the scientific arrangements producing the answers.

The first half of the 20th century is also relevant to this report for what it did not establish. There was nothing established in the social sciences comparable to the National Institutes of Health, and there were no government laboratories akin to Lawrence Livermore or Los Alamos. The basic arrangement in the biological and physical sciences includes government conducted and managed research along with independent, university-based research and corporate research, especially in medicine and electronics. This did not develop for the social sciences. Rather, the model that emerged situated the production of social science knowledge primarily in the nonprofit sector and funded from a mixture of private and public sources. And though there are instances of industry-based advances in social science methods—survey research and psychological testing being the leading examples—these occur much less frequently in the social than in the natural sciences. The basic positioning of the social sciences in the nonprofit sector has implications for its use in policy making, notably in the number and workings of intermediary organizations—think tanks and advocacy organizations—and in the heavy presence of interested private funding.

CHANGING PERCEPTIONS OF USE

Scope of the Policy Enterprise

The postwar era of big science introduced new challenges to understanding the usefulness of the social sciences. This is first evident in the sheer scope of the enterprise. In the 1920s, there were a handful of privately funded, nongovernmental social science organizations—the National Bureau of Economic Research, the Social Science Research Council, and the Brookings Institution, are notable examples—established in part to offer expertise and advice to the government. There was growth in the World War II era and shortly thereafter—the RAND Corporation, the American Institutes for Research, ISR and NORC, for example—and, importantly, federal funds were added to what had largely been philanthropic funding. The request-for-proposal (RFP) mechanism, initially designed to purchase military hardware, was re-engineered to purchase scientific expertise.[4] For-profit and nonprofit research contract houses were formed, and consulting firms got into the business of using social science to advise government and private clients. A recent survey estimates more than 1,800 think-tank-like

---

[4]For example, NORC, a midsize social science organization with an annual budget of approximately \$160 million, annually screens about 10,000 government-issued RFPs, closely examines 10 to 15 percent of them, and prepares formal bids for several hundred.

organizations in the United States (McGann, 2010), approximately a quarter of which are based in Washington, DC.

Another big change occurred in higher education. When big social science got under way in the 1960s, public policy schools were few; today there are several hundred master's level degree programs in public administration or public policy.[5] Graduates from these programs fill the growing number of positions available in policy analysis and advocacy institutions.[6] By any measure—absolute numbers, budgets, number of employees, available federal funds, research output, scope of policies addressed—the current nongovernmental organization of policy analysis, advice, and advocacy is vastly different from that which characterized the first half of the 20th century. It does not exaggerate to label what is now in place a policy enterprise, which we can now define as an interlocking array of institutions and practices that use (or claim to use) science to influence policy making. Funds come from private and public sources. Influence flows through informal briefings, publications, media placement, as well as more formal arrangements such as RFPs and consultancies. There is continual circulation of personnel among the institutions that make up this enterprise, as well as circulation in and out of government positions.

A New Focus on Use

One early feature of the policy enterprise was a new research specialty dedicated to studying how research knowledge is used. It began in earnest in the work of Cohen and Lindblom (1979) and Weiss (1977), and in the scholarship of Campbell and his colleagues on social experimentation and the role it should play in shaping public policy (e.g., Campbell, 1975; Cohen and Garet, 1975; Dehue, 2001; Floden and Weiner, 1978; Riecken and Boruch, 1975). In this literature—spanning the fields of sociology, organizational behavior, political science, psychology, education, and, more recently, science and technology studies—understanding the use of social

---

[5]This number includes accredited graduate degree programs and graduate schools in the field of public administration and policy at the master's levels in the United States from all types of institutions except online degrees: see http://www.gradschools.com/programs/public-affairs-policy [July 2012].

[6]For example, the Research Triangle Institute has a staff of approximately 2,800 people (see http://www.rti.org [January 2012]); Westat, more than 2,000 (see http://www.westat.com [January 2012]); RAND Corporation, approximately 1,600 (see http://www.rand.org [January 2012]); and the American Institutes for Research, 1,500 (see http://www.air.org [January 2012]).

science in public policy moved to the foreground. An institutional expression of this interest was the Center for the Research Utilization of Scientific Knowledge (CRUSK), established in 1970 by the Institute for Social Research at the University of Michigan. Other ISR centers established in this period flourished and continue today to play major roles in social science. The center for research utilization, however, did not flourish. It was closed in 1985, an early indicator of how difficult it is to study the phenomenon of use (Frantilla, 1998).

A key document of the period was a 1978 National Research Council (NRC) report, *Knowledge and Policy: The Uncertain Connection.* More than three decades later we find that its major conclusion strikingly anticipates one we reach. The 1978 report made clear that the question of use had ceased to be primarily one debated within social science as was the case in the prewar decades. By the late 1970s, the use question engaged a broad community of potential users and intermediaries, as well as academic researchers. The report (National Research Council, 1978, p. 1) opened with a worry:

> Although the need for large-scale federal support of social R&D [research and development] is widely accepted, questions concerning its relevance to the making of social policy have become more insistent in recent years. What are we learning? Who is making effective use of what we learn?

The report traced the questioning to two sources: "legislators distrustful of 'social engineers' who promote radical ideas or pursue irrelevant academic interests, and social scientists worried that dependence on government might compromise their objectivity" (p. 2). Although echoes of these points can be found in today's discussions, they are not central to how the usefulness of social science is framed in this report.

Rather, we start with another observation in the NRC report: "the policy world now [1970s] takes it for granted that the social sciences have a contribution to make in government" (p. 4). The report listed numerous innovations that were designed to reduce the uncertainty in the connection between knowledge and policy. They included competitively awarded contracts, collaboration between funders and recipients of funds, and program evaluation.

But the report then reached a sobering conclusion (National Research Council, 1978, p. 5):

> Unfortunately, we lack systematic evidence as to whether these steps are having the results their sponsors hope for. . . . [S]ocial R&D

continues to be criticized by members of Congress, executive-branch officials, and social scientists because it is neither good nor well-managed research and has little potential for use.

The report continues in this vein, asking: "What knowledge do we possess that is relevant to the formulation of social R&D policy? Regrettably (and ironically), we possess little knowledge obtained through research that will help answer [this] question" (p. 6).

In the 35 years since that NRC report, the policy enterprise concerned with bringing social science to bear on policy making has steadily expanded, received more funding, and become more professional. However, the telling conclusion just cited—"we lack systematic evidence as to whether these steps [to connect scientific knowledge and policy] are having the results their sponsors hope for"—is one we reach today.

Our explanation for this is simple. Like our predecessor committee, we take for granted that using science contributes to policies that are more likely to result in consequences that policy makers intend. Today, however, we conclude that in the years since the 1978 report the focus on operationalizing "use" has not provided an adequate understanding of what happens between science and policy in policy making, a point developed in Chapter 3.

SCOPE OF INVESTMENT IN THE POLICY ENTERPRISE

The nation invests in the social sciences and, even if not in amounts characteristic of physics, biology, or chemistry, still at nontrivial levels. The National Science Foundation (2012a, Table 19) estimates that federal government obligations for social science research in fiscal 2011 were $1.3 billion and that its own expenditures for social, behavioral, and economic sciences in fiscal 2012 would be $254 million (National Science Foundation, 2012b). Many other agencies, such as the Environmental Protection Agency, the Food and Drug Administration, the Central Intelligence Agency, and the Departments of Defense, Education, and Health and Human Services (HHS),[7] also fund social science research. In addition to these direct expenditures, there are federal data collection activities that generate

---

[7]The National Institutes of Health (2012a), a part of HHS, reports for fiscal 2013 that about 10 percent of its $30 billion budget is to fund behavioral and social science research, including projects that involve social scientists working with biological and medical scientists (Silver et al., 2012).

social and economic statistics used extensively by social scientists investigating a very wide range of policy questions. The fiscal 2012 budgets for the 13 principal statistical agencies designated by the U.S. Office of Management and Budget (OMB)[8] exceeded $2.6 billion, not including $498 million in that year for the decennial census. Other federal statistical programs added $3.6 billion to the generation of social and economic data (U.S. Office of Management and Budget, 2011). Significant investments from the private sector foundations, corporations, and individuals join these public funds in supporting research universities and policy institutions.

From these public and private (generally tax-exempt) sources, there is a several billion dollar investment in social science research. That investment includes support for scientists working in universities, research institutions, and linkage institutions, as well as those in government agencies—on a large scale in intelligence, national security, and defense agencies and in statistical agencies (particularly the Census Bureau and the Bureau of Labor Statistics) and, on a numerically smaller though still influential scale, in legislative and executive offices, such as the Congressional Budget Office and the HHS Office of the Assistant Secretary for Planning and Evaluation. Most states and many local governments duplicate features of these programs and offices. The number of social science trained experts working in governments across the country reaches into the several thousands.

One standard justification for government-subsidized social science is that it produces a public good in the form of reliable and credible knowledge for policy making that would not otherwise be produced. But of course

---

[8]The 13 principal statistical agencies are the Bureau of Economic Analysis and the Bureau of the Census in the Department of Commerce; the Bureau of Justice Statistics of the Department of Justice; the Bureau of Labor Statistics of the Department of Labor; the Bureau of Transportation Statistics in the Research and Innovative Technology Administration of the Department of Transportation; the Economic Research Service and the National Agricultural Statistics Service of the Department of Agriculture; the Energy Information Administration of the Department of Energy; the National Center for Education Statistics in the Institute of Education Sciences of the Department of Education; the National Center for Health Statistics in the Centers for Disease Control and Prevention of the Department of Health and Human Services; the National Center for Science and Engineering Statistics in the Social, Behavioral and Economic Sciences Directorate of the National Science Foundation; the Office of Research, Evaluation, and Statistics of the Social Security Administration; and the Statistics of Income Division in the Internal Revenue Service Office of Research, Analysis, and Statistics of the Department of the Treasury. These agencies constitute 13 of the 14 members of the Interagency Council on Statistical Policy. The 14th member is the Office of Environmental Information in the Environmental Protection Agency, which is not a self-contained statistical agency (National Research Council, 2009).

the value of this effort depends on the science being used. A starting point, then, is asking what is known about use. The next chapter explains why scholarship on use to date is inadequate. Chapter 4 follows with a framework for research that can extend and deepen the understanding of use.

3

The Use of Research Knowledge: Current Scholarship

With the arrival of big social science and the growth of the policy enterprise, the federal investment in social science brought attention to whether the knowledge being produced was being used. Research on what was labeled "knowledge utilization" got under way. We address that research under three headings: decisionism and its critique, the metaphor of two communities (researchers and policy makers), and the evidence-based policy and practice initiative.

As an introduction to these issues we take brief note of the characteristics of our three central topics—social science, policy, using science—that challenge any attempt at a comprehensive account for the when, how, and why of science use in policy.

A CHALLENGING LANDSCAPE

Scholarship on what happens at the interface of science and policy has to contend with two phenomena—policy making and use—that are particularly difficult to define. To begin with, investigations of these phenomena are launched in different disciplines, including anthropology, political science, psychology, and sociology and their myriad subfields and cross-fields, from science and technology studies to political psychology, from behavioral economics to historical sociology. Each of these fields has its own established principles of evidence and inference. They use different methods—experimental, analytic, quantitative, and qualitative. They work at different levels of analysis—from individual behavioral decision theory to systems theory. They focus on different processes: from structural determinism and constrained probabilities at one end of a continuum to willful effort and chance happenings at the other. They draw on epistemologies as varied as positivism, critical realism, and postmodernism. Individual social scientists bring different motivations to their work—from expansion of theoretical knowledge to practical problem solving, from mapping policy options to advocacy of particular policies. Social scientists bring their expertise to universities, think tanks, the media, advocacy groups, corporations, and government agencies. This range—across fields of study and individual motivations and career lines—produces a lot of variability, which, of course, determines the way the science-policy nexus is framed.

Complicating matters is the absence of a generally accepted explanatory model of policy making. Instead, multiple descriptive policy process models offer ways to understand how policy is made and how science might enter into that process. There are, for example, rational models—including linear, cycle or stage, incrementalism, and interactive. There are models that question rational model assumptions, including behavioral economics, path dependency, and bureaucratic inertia. There are political models, including policy networks, agenda setting, policy narratives, advocacy coalition frameworks, punctuated equilibrium theory, and deliberative analysis models (see Baumgartner and Jones, 1993; Hajer and Wagenaar 2003; Kingdon, 1984; Lindblom, 1968; Neilson, 2001; Sabatier, 2007; Sabatier and Jenkins-Smith, 1993; Stone, Maxwell, and Keating, 2001).

There are models that focus on different stages of the policy process and thus on different ways that social science can contribute, including: descriptive analyses that present conditions needing policy attention, such as a slowdown in small business start-ups; social indicators that document long-term trends, such as gender differences in pay scales; social experiments on alternative policy designs, such as school vouchers; and evaluation research on the effectiveness of a policy, such as neighborhood policing.[1]

Political science is the discipline that has devoted the most attention to the policy process. On the issue of use, it has reached a general conclusion (Henig, in press):

---

[1] For a careful discussion of how evidence is used at different stages of the policy process, see McDonnell and Weatherford (2012).

[T]he main thrust of the political science literature serves as a warning against idealized visions of pure data being applied in depoliticized arenas. Although generalizations about an entire discipline inevitably are oversimplifications, the center of gravity within the field encourages skepticism about proposals for a rational, comprehensive, science of public policy making and regards data and information as sources of power first and foremost.

It is difficult to assess how widely this characterization is accepted outside of political science, but it is clear that the various models and frameworks do not coalesce into anything remotely resembling a powerfully predictive, coherent theory of policy making. Lacking that, it is improbable and perhaps impossible to reach a widely agreed-upon understanding of the use of science in policy making. "Use" itself, consequently, is elusive, seen differently depending on the perspectives brought to it and the policy and institutional arenas in which it is investigated (Neilson, 2001; Webber, 1991; Weiss, 1991). A political psychologist at the Central Intelligence Agency concerned with what transforms an angry, unemployed teenager into a terrorist uses research evidence very differently from an economist at the RAND Corporation designing a randomized controlled field trial (RCFT) on classroom size and school performance. Many researchers underscore the conceptual confusion about use and conclude that different definitions of use are needed and appropriate for different purposes (e.g., Oh, 1997; Rich, 1997; Weiss, 1979).

This conclusion is consistent with the fact that policy choices are context dependent. A school district deciding whether to establish charter schools is less interested in a comparative study of charter and public schools across the country than in wanting to know how well a charter school will perform under its conditions, which differ depending on whether the district is in the central city or suburb, with a homogenous or diverse population, with a historically competent or incompetent school administration. The usefulness of research is not assessed in terms of variance explained from a large sample of schools, but whether it is informative about a very specific choice.

Given the context-dependent nature of the use of science, typologies are a common way of mapping the landscape (for a summary, see Nutley et al., 2007; see also Bogenschneider and Corbett, 2010; Renn, 1995). A frequently cited typology is that of Weiss (1979, 1998; see also Weiss et al., 2005):

- *Instrumental uses* occur when research knowledge is directly applied to decision making to address particular problems.
- *Conceptual uses* occur when research influences or informs how policy makers and practitioners think about issues, problems, or potential solutions.
- *Tactical uses* involve strategic and symbolic actions, such as calling on research evidence to support or challenge a specific idea or program, such as a legislative proposal or a reform effort.
- *Imposed uses* (which is perhaps a variant on instrumental uses) describe mandates to apply research knowledge, such as a requirement that government budgeting be based on whether agencies have adopted programs backed by evidence.

Other scholars add a fifth category, symbolic or ritualistic use—that is, the organizational practice of collecting information with no real intent to take it seriously, except to persuade others of a predetermined position or even to delay action (Leviton and Hughes, 1981; Shulha and Cousins, 1997). It is a frequent complaint among scientists that policy makers use scientific evidence as confirmation of prior beliefs. This complaint, however, overlooks the fact that, when policy makers argue on the basis of evidence, it is more difficult for their opponents to ignore that evidence, or to leave it unchallenged. "My science versus your science" has the merit of putting science in play, and over time opens more space for policy arguments that include scientific evidence.

Weiss emphasizes that each of the four uses—which also applies to the fifth use noted—can be found in particular situations, but that no one of them offers a complete picture. Scholars who debate typologies of use generally conclude that, although typologies are heuristically valuable, they are not easily applied empirically. Boundaries are blurred, and access to users' cognitive processes is unattainable. In fact, it is unlikely that users themselves can make sharp distinctions in explaining how they use knowledge (Contandriopoulos et al., 2010). The empirical application of typologies in research is difficult because use is "a dynamic, complex and mediated process, which is shaped by formal and informal structures, by multiple actors and bodies of knowledge, and by the relationships and play of politics and power that run through the wider policy context" (Nutley et al., 2007, p. 111).

Typologies of use fail to meet the standard criteria of scientific typologies in which each category consists of an internally coherent set of variables, with the value of each variable predictably correlating with the values of each of the other variables in that particular category. In the periodic table of chemical elements, for example, hydrogen is distinguished from other chemical elements by its atomic weight, its specific gravity, its bonding properties, the temperature at which it freezes and boils, and other traits. Each of these traits differs consistently and predictably from those same traits in helium or in any other chemical element (see Stinchcombe, 1987). In the social world it is impossible, in any practical sense, to construct typologies that meet this standard. Typologies of social conflict, ethnic or racial groups, or government corruption are never going to have categories with internally coherent variables whose values covary in completely predictable ways. It is unrealistic to expect a clear and unambiguous typology for a phenomenon as complex as the use of science in policy.

To address the charge given to this committee—to understand the use of science in policy—is thus to simultaneously deal with three elusive phenomena:

- Scientific findings from multiple sources and that are at times contradictory;
- A policy-making process, that is variable along many dimensions; and
- A phenomenon, "use," that changes its meaning depending on the perspective brought to it and one's location in the complex space where policy is made.

With this challenging landscape in mind, we turn to the recent scholarship on knowledge utilization.

DECISIONISM AND ITS CRITIQUE

The scholarship on knowledge utilization has, virtually from its beginnings, been skeptical of rational models of the relationship between research and policy. Rational models assume that decisions unfold through five stages (Nutley and Webb, 2000, p. 25):

1. A policy problem requiring action is identified and goals, values, and objectives are clearly set forth;
2. All significant ways of addressing the problem and achieving the goals or objectives are enumerated;

3. The consequences of each alternative are predicted;
4. The consequences are then compared with the goals and objectives; and
5. A strategy is selected in which consequences most closely match the goals and objectives.

Weiss and Bucuvalas (1980, p. 263) summarized the essence of this model: "a decision is pending, research provides information that is lacking, and with the information in hand the decision maker makes a decision." Rational models have also been characterized as "decisionism"—"a limited number of political actors engaged in making calculated choices among clearly conceived alternatives" (Majone, 1989, p. 12; see also Rein and White, 1977; Rich, 1997).

Criticisms of this model have focused on several significant defects; for example, that decisions made are optimal, that is, based on complete information and an examination of all possible alternative courses of action (see the work of Simon [1957], who introduced satisficing as a replacement for maximizing); or, that the model is a normative account of policy making (see the work of Braybrooke and Lindblom [1963] and Lindblom [1959], authors who substitute incrementalism for rational models). Other critics argue that rational models underemphasize or ignore the important role that value judgments play in policy arguments (Brewer and deLeon, 1983); or that linear problem solving is "wildly optimistic," because it "takes an extraordinary concatenation of circumstances for research to influence policy decisions directly" (Weiss, 1979, p. 428).

More recent examinations of the relationship between research and policy making echo these concerns. For example, Gormley (2011, pp. 978-979) notes:

A hypodermic needle theory of scientific impact on policy, which anticipates direct, immediate, and powerful effects, is flawed for several reasons. First, scientific research is one of many inputs into the policy process. . . . Second, scientific knowledge accumulates through multiple studies, some of which reach different conclusions. . . . Third, the applicability of a given study to a particular policy choice is a matter of judgment. . . . Fourth, scientific research is translated, condensed, repackaged, and reinterpreted before it is used. Fifth, the use of scientific information by public officials, when it is occurs, is more likely to involve justification (reinforcement of a prior opinion) than persuasion (conversion to a new opinion).

Although we share Gormley's view, there are situations in which discrete decisions are directly triggered by the use of some specific scientific knowledge—for example, the direct, even formulaic translation of census results into congressional apportionment or formula-based fund allocations that are legislatively required. There also are situations in which a user is considered sovereign in her or his capacity to mobilize evidence and, consequently, to modify her or his behavior on the basis of that evidence—for example, the choice of a preferred clinical treatment (Contandriopoulos et al., 2010). But these examples are exceptions to the rule, and uncommon at that. It is estimated that evidence-based programs accounted for less than 0.2 percent of nonmilitary discretionary spending in fiscal 2011.[2]

In almost all decision-making situations, the use of science takes place in "systems characterized by high levels of interdependency and interconnectedness among participants" (Contandriopoulos et al., 2010, p. 447). No single decision maker has the independent power to translate and apply research knowledge. Rather, multiple decision makers are embedded in systemic relations in which use not only depends on the available information, but also involves coalition building, rhetoric and persuasion, accommodation of conflicting values, and others' expectations.

In criticizing rational models and decisionist thinking, Weiss and others suggest that use is less a matter of straightforward application of scientific findings to discrete decisions and more a matter of framing issues or influencing debate (Weiss, 1978, p. 77):

> Social science research does not so much solve problems as provide an intellectual setting of concepts, propositions, orientations and empirical generalizations. . . . Over a span of time and much research, ideas . . . filter into the consciousness of policy-making officials and attentive publics. They come to play a part in how policy makers define problems and the options they examine for coping with them.

---

[2]The George W. Bush administration piloted a program linking federal financing to clear demonstration of program effectiveness. These evidence-based programs "accounted for about $1.2 billion out of a $670 billion budget for nonmilitary discretionary programs in the 2011 fiscal year" (Lowrey, 2011).

Although Weiss suggested that this enlightenment model is perhaps the way science is most frequently used in policy making, she did not claim it was the way it *ought* to happen. "Many of the social science understandings that gain currency are partial, oversimplified, inadequate, or wrong. . . . The indirect diffusion process is vulnerable to oversimplification and distortion, and it may come to resemble 'endarkenment' as much as enlightenment" (Weiss, 1979, p. 430).

In sum, the research on knowledge utilization reflects a consensus about what should be ruled out: (1) that the science/policy nexus can be uniformly understood in terms of rational decision-making models; (2) the assumption of a specified single actor with freedom to achieve goals formulated through a careful process of rational analysis characterized by a complete, objective study of all relevant information and options; and (3) the definition of use as problem solving in the sense of a direct application of evidence from a specific set of studies to a pending decision. Although evidence may occasionally be used in such narrow ways, these depictions of "use" do not accurately reflect the full realities of policy making.

Knowledge utilization research, in agreement about what is ruled out, is less clear about what should be ruled in. It has, however, pointed to the importance of closing the distance between the "two communities" of scientists and policy makers.

THE TWO COMMUNITIES METAPHOR

Viewing use from the perspective of two communities has been a recurring motif in knowledge utilization studies (see Caplan, 1979). The basic idea is refreshingly simple. Scientists and policy makers are separated by their languages, values, norms, reward systems, and social and professional affiliations. The primary goal of scientists is the systematic search for a reliable and accurate understanding of the world; the primary goal of policy makers is a practical response to a particular public policy issue.

Like any binary distinction, this one oversimplifies, though there is a crude truth to several distinctions rooted in the different tasks facing researchers and policy makers. They differ in the outcomes they value—knowledge about the world in all its complexities versus knowledge helpful in reaching feasible solutions to pressing problems—and in the incentives, rewards, and cultural assumptions associated with these different outcomes. They also differ in habits of expression—probabilistic versus certain statements about conditions or people. And they differ even in modes of thought—deductive and general versus inductive and particular (Szanton, 2001, p. 64). This is described as "research think" and "political think." The "culture of the researcher tends to add complexity and resist closure. The culture of the political actor tends to demand straightforward and easily communicated lessons that will lead to some kind of action" (Henig, 2009, p. 144).

Differences between the two communities are associated with a contrasting list of supply-side and demand-side problems (Bogenschneider and Corbett, 2010; Furhman, 1994; Nutley et al., 2007; Rosenblatt and Tseng, 2010). On the supply side are researchers who fail to focus on policy-relevant issues and problems, cannot deliver research in the time frame generally necessary for effective policy making, do not relate findings from specific studies to the broad context of a policy issue, ineffectively communicate their findings, depend on technical arguments that are inaccessible to policy makers, and lack credibility because of perceived career interests or even partisan biases. On the demand side are policy makers who fail to spell out objectives in researchable terms, have few incentives to use science, and do not take time to understand research findings relevant to pending policy choices.

This framing of the use problem offers little guidance as to which of the long list of factors, from either side, best explains variance in use, let alone how the factors interact and whether they apply only in specific settings or have general applicability (Bogenschneider and Corbett, 2010; Johnson et al., 2009). Although the two communities framework has been helpful in understanding the differing expectations of researchers and policy makers and problems of communication between them, it has not been able to offer a systematic explanation of use. Thinking about how best to bridge the gap between the two communities has, however, led to practices of translation and brokering and to more intensive interactions between researchers and policy makers.

Translation

Translation is a supply-side solution to the use problem. It was developed in clinical diagnostic, preventive, and therapeutic practices. The idea is simple: basic science is translated into clinical efficacy, efficacy is translated into clinical effectiveness, and effectiveness is translated into everyday health care delivery (Drolet and Lorenzi, 2011). The oft-invoked catchphrase is "bench to bedside." One important sign of the seriousness with which translation is taken is the U.S. Department of Health and Human Services initiative, Translating Research into Practice (TRIP) Program, that focuses on implementation techniques and factors associated with successfully translating research findings into diverse applied health care settings (see Agency for Healthcare Research and Quality, 2012).

Translational strategies have now moved beyond health care, introducing additional and somewhat differently focused activities. One is evidence-based registries, a compilation of scientifically proven interventions. They are considered tools to improve practice in various fields, including social services, criminal justice, and education. A different initiative is the Campbell Collaboration,[3] an international organization conducting systematic reviews of the effects of social interventions.

The translation strategy is well institutionalized in education. The U.S. Department of Education's Institute of Education Sciences (IES) was established in part to develop the science that could be translated into strategies to change education practice in public schools. The What Works Clearinghouse of the IES aims to provide educators, policy makers, and the public with an independent, and trusted source of scientific knowledge relevant to education policies and practices.[4] IES also supports 10 regional educational laboratories, the role of which is similar to that of extension agents in the agricultural field: taking research results and putting them into practice in school districts and classrooms (see U.S. Department of Education, 2012).

The movement toward evidence-based approaches in practice settings began more than 40 years ago in medical practice. Archibald Cochrane (1972) railed against ineffective and sometimes harmful therapies despite randomized clinical trials showing that better treatments were available. In response to his call for systematic reviews of such trials, the Cochrane Collaboration[5] was established. Its rigorous model of research synthesis has been adopted in other fields, including the above-noted Campbell Collaboration and the What Works Clearinghouse.

Although translation strategies have largely been applied to practices, the logic of translation is applicable to questions of using science in policy. Begin with a dependable, valid scientific base that provides evidence about

---

[3]See the Campbell Collaboration: What Helps? What Harms? Based on What Evidence?, available: http://www.campbellcollaboration.org/ [August 2012].

[4]For example, see the IES guides in education, such as "Turning Around Chronically Low-Performing Schools" (May 2008): available: http://ies.ed.gov/ncee/wwc/practiceguide.aspx?sid=7 [July 2012].

[5]See the Cochrane Collaboration: available: http://www.cochrane.org/index.htm [August 2012].

what works so that policy makers can readily grasp its relevance to the decision or task at hand, and make that science available in the form of research summaries or lists of demonstrably effective social interventions. The research record, however, is far from clear on whether translation (of either social or medical science research) works and is an effective strategy for enhancing use (see, e.g., Glasgow and Emmons, 2007; Green and Seifert, 2005; Lavis, 2006; Slavin, 2006).

Brokering

While translation is primarily a matter of repackaging technical findings in terms more readily consumable by policy makers, brokering is a two-way conversation aided or mediated by a third party. Brokering involves filtering, synthesizing, summarizing, and disseminating research findings in user-friendly packages. It is generally seen as the task of intermediary organizations, such as think tanks, evaluation firms, and policy-oriented organizations, including those focusing on specific target populations or specific social issues as well as those organized around particular political persuasions. These organizations (Bogenschneider and Corbett, 2010, p. 94):

> do research and evaluation, but they also have one foot in the policy world. They see policymakers as their primary clients. In addition to producing knowledge, they also see their role as translating extant research and analysis in ways that enhance their utility for those doing public policy. . . . To greater and lesser degrees, these firms bridge the knowledge-producing and knowledge-consuming worlds.

Science and technology studies describe brokering as occurring in boundary organizations occupying a territory between research and policy making (Guston, 2000).[6] In contrast to translation strategies that generally are one-way efforts in dissemination, brokering involves interaction and two-way communication. Intermediary organizations and knowledge brokers are increasingly being viewed as critical in promoting the capacity for evidence-based, or evidence-informed, decision making (e.g., Dobbins et al., 2009a).

---

[6]In this view, the National Research Council can be viewed as a brokering organization, synthesizing research in a consensus-based process and then presenting it in a form intended to contribute to improved policy making.

If brokering occurs, use is not something that happens when experts "here" hand off research to policy makers "there." A brokering model views use as emerging from multidirectional communication and ongoing negotiation among researchers, policy makers, planners, managers, service providers, and even the public. Often this interactive process will involve consideration of more than one stream of research as relevant to a given policy (e.g., Sudsawad, 2007).

To bridge the gap between the differing cultures of the producers and consumers of scientific knowledge will require, according to some scholars, cultural changes in each community. Bogenschneider and Corbett (2010, pp. 299 ff.) write that the culture of research should change, perhaps through education and training on how to do more policy-relevant research, developing incentives for doing such research and developing opportunities to work with policy makers. The user or consumer culture should also change, perhaps by institutional innovations that improve policy makers' access to research, helping them communicate their policy needs to researchers, and providing forums to discuss research agendas. In more ambitious formulations, research literacy of the general public should be improved through education (see also Carr et al., 2007; Gigerenzer et al., 2008).

An Interaction Model

Closing the distance between the two communities has taken an additional step in what is labeled the interaction model (Contandriopoulos et al., 2010; Greenhalgh et al., 2004). This model goes beyond transfer, diffusion, and dissemination and even beyond translation and brokering. The interaction label covers a family of ideas directed to systemic changes in the means and opportunities for relationships between researchers and policy makers (Bogenschneider and Corbett, 2010). It holds that the relation between researchers and users is not only not linear it is iterative and even "disorderly" (Landry et al., 2001, p. 335).

One source for an interest in interaction is science and technology studies documenting the co-evolution of social and technological systems (Jasanoff, 2004; Jasanoff et al., 1995). Another source is the use of systems thinking to better understand the complex adaptive systems involved in diagnosing and solving public health problems and the interactions among the design of prevention interventions, testing their efficacy and effectiveness, and disseminating innovations in community practices. A third is the emphasis on practical reasoning, the argumentative turn in policy analysis discussed in the next chapter (Fischer and Forester, 1993; Hajer and Wagenaar, 2003; Hoppe, 1999).

Research that works in close proximity to practice settings illustrates the interaction framework. First noted in corporate research (Pelz and Andrews, 1976), and later in the life sciences (Louis et al., 1989), the publication of *Pasteur's Quadrant* (Stokes, 1997), with its emphasis on use-inspired research, increased its visibility. This research influenced how the National Academy of Education (1999) set research priorities and its interest in how to hold policy specialists, researchers, and professional educators, program developers, and curriculum specialists collectively accountable for educational outcomes. Collaborations of this kind formed the basic design concept for the Strategic Education Research Partnership. These involved connecting researchers to teachers, bringing in research communities, school administrations, and educational policy makers (see National Research Council, 1999a; Smith and Smith, 2009). The Carnegie Foundation for the Advancement of Teaching and Learning is also promoting a framework for research and development labeled improvement research (Bryk et al., 2011), which synthesizes the work of researchers and practitioners.

In this spirit, the Institute of Medicine (IOM) created a Roundtable on Evidence-Based Medicine, which then became the Roundtable on Value & Science-Driven Health Care, to foster interaction among stakeholders interested in building a continuously learning health care system in which science, information technology, incentives, and culture are aligned to bring together evidence-based practice and practice-based evidence (see Green, 2006). This effort and its attendant workshops (Institute of Medicine, 2007, 2010b, 2011a, 2011b) stress the importance of rigorous science and applying the best evidence available. The goal is understanding how health care can be restructured to develop knowledge from science and from the health care process and to then apply it on many fronts: health care delivery and health improvement, patient and public engagement, health professional training, infrastructure development, measurement, costs and incentives, and policy. The IOM's reports on these activities draw attention to active collaboration, exchange, and appraisal of research and policy and to what is known by researchers and users of research about practice—drawn from the life-cycle of therapies, their development, testing, introduction, and evaluation.

As attractive as these initiatives are, there are cautionary voices. There is a difference across political time, policy time, and research time. One should take care not to mistake one for another (Henig 2009, p. 153):

The pressure for fast, simple, and confident conclusions, however, is generated by the needs of politicians—not necessarily the needs of the policy. Political time is defined by election cycles, scheduled reauthorization debates, and the need to respond to short-term crises or sudden shifts in public attention. But a consideration of the history of public policy suggests that societal learning about complex problems and large-scale policy responses takes place on a much more gradual curve.

Interaction models offer an insight into what the use of science means in practice. Evidence from science is not simply there for the taking. It emerges and is made sense of in the particular circumstances that give rise to a policy argument (see Chapter 4 for discussion of policy argument). "Making sense" is iterative. It involves negotiating what kind of situation-specific knowledge is relevant to a policy choice, whether it is firmly established and available under the constraints of time and budget, and what political consequences might follow from using it. In this framework, formal linkages and frequent exchanges among researchers, policy makers, and service providers occur at all steps between knowledge production and knowledge use (Huberman and Cox, 1990). What emerges is a *social* as well as a technical exercise. Conklin et al. (2008, p. 7) explain this framework:

> Strategic interactions (between human actors within and between organizations) therefore address both sides of the research-policy interface. On the one hand, decision-makers highlight policy relevant research priorities; on the other hand, researchers can interpret research findings in local contexts. In so doing, a common understanding of a policy problem, and its possible solutions, is built between different actors in the two communities. . . .

Spillane and Miele (2007) underscore the point in observing that *what* information is noticed in a particular decision-making environment, *whether* it is understood as evidence pertaining to some problem, and *how* it is eventually used all depend on the cognitions of the individuals operating in that environment. Furthermore, what these actors notice and make sense of is determined in part by the circumstances of their practice environment. Examining use, then, also requires examining "the practice of sense making, viewing it as distributed across an interactive web of actors and key aspects of their situation—including tools and organizational routines"

(p. 49). It also introduces the idea that research might "be interpreted and reconstructed—alongside other forms of knowledge—in the process of its use" (Nutley et al., 2007, p. 304).

Focusing on understanding institutional arrangements—how the agencies, departments, and political institutions involved in policy making operate and relate to one another—may be what matters most in improving the connection between science and policy making. For example, a study of drug misuse in government agencies in Scotland and England (Nutley et al., 2002) suggests that three aspects of microinstitutional arrangements within and between the agencies mattered a great deal in understanding how research evidence was (or was not) used:

1. How different agencies integrated research with other forms of evidence,
2. How agencies collectively dealt with the fragmentation of research evidence resulting from different agencies producing different types of evidence given their respective research cultures, and
3. What mechanisms were in place to integrate evidence and policy making (co-location of research and policy staff, cross-government work groups, establishment of quasi-policy bodies that specialize in the substance of a policy domain, etc.)?

Nutley et al. (2007, pp. 319-320) conclude

[T]here is now at least some credible evidence to underpin [their view] . . . that interactive, social, and interpretive models of research use—models that acknowledge and engage with context, models that admit roles for other types of knowledge, and models that see research use being more than just about individual behavior—are more likely to help us when it comes to understanding how research actually gets used, and to assist us in intervening to get research used more. . . .

If this conclusion holds up, it is a step toward accumulating what the committee believes is lacking: understanding institutional arrangements that facilitate the use of science in policy.

There is an important cautionary observation about efforts to overcome the "two communities" challenge. There are tensions between scientific engagement with practical policy problems and the long-standing assumption that science maintains its authority by virtue of its independence from politics (Jasanoff, 1990; Jasanoff et al., 1995). Persons working to bring scientists and policy makers closer need to be mindful that this tension is never far from how scientists think about and engage the policy uses of their work.

EVIDENCE-BASED POLICY AND PRACTICE

Current discussions about the use of research knowledge are heavily influenced by "evidence-based policy and practice." The goal is realizing better and more defensible policy decisions by grounding them in the conscientious, explicit, and judicious use of the best available scientific evidence (Davies et al., 2000). The initiative explicitly rejects habit, tradition, ideology, and personal experience as a basis for policy choices: they are to be replaced with a more dependable foundation of "what works," that is, what the evidence shows about the consequences of a proposed policy or practice. With access to an evidence base, argue the proponents, policy makers will make better decisions about the direction, adoption, continuation, modification, or termination of policies and practices. Dunworth et al. (2008, p. 7) note:

> [W]hile scientific evidence cannot help solve every problem or fix every program, it can illuminate the path to more effective public policy. . . . [T]he costs and lost opportunities of running public programs without rigorous monitoring and disinterested evaluation are high . . . without objective measurements of reach, impact, cost effectiveness, and unplanned side effects, how can government know when it's time to pull the plug, regroup, or, in business lingo, "ramp up?"

The use of science is, of course, not a logical or inevitable outcome of having the science. In fact, the normative claim that policy should be grounded in an evidence base "is itself based on surprisingly weak evidence" (Sutherland et al., 2012, p. 4).

The approach of evidence-based policy and practice assumes that there is an agreement among policy makers and researchers on what the desired ends of policy should be. "The main contribution of social science research is to help identify and select the appropriate means to reach the goal" (Weiss 1979, p. 427). This, in turn, depends on the quality of the science providing evidence to the policy maker, and thus the evidence-based approach places a premium on improving policy-relevant research, often through the use of RCFTs.

In the settings in which they are carried out, RCFTs provide a strong, if not the strongest, form of scientific evidence of cause and effect. Circumstances may permit such experiments in a desired setting, such as when scarce resources are allocated by lottery, for example with admission to magnet schools or charter schools or the allocation of health care resources. An example of the latter is the Oregon Health Insurance Experiment in which names were drawn by lottery for the state's Medicaid program for low-income, uninsured adults (Finkelstein et al., 2012).

Even when RCFTs are conducted in one setting, inference from them may be applied to other settings or contexts with concurrent collection of information on other variables or factors that differ in different settings and that may influence the results. So-called substitutes for randomized trials, however, such as "natural" experiments and "quasi-experiments," as Sims (2010) argues, are not actually experiments. They are often invoked as a way to avoid confronting "the complexities and ambiguities that inevitably arise in nonexperimental inference." For these situations and even in conjunction with randomized experiments, there are nonexperimental methods of drawing causal inferences and model-based methods for adjusting experimental results for inherent biases. Appendix A provides a review of some of these research methods and sets them in the context of the varied statistical methods for research and evaluation.

The active debate regarding the appropriate methodology for a given research question promotes attention in the policy community to the desirability of producing the best possible evidence under a given set of circumstances, especially the strongest evidence that bears on policy implementation and policy consequences. Bringing attention to the importance of strong evidence in policy making advances the goal of using science even though the specific formulation of an evidence-based policy approach offers little insight into the conditions that bring about its use.

CONCLUSION

Despite their considerable value in other respects, studies of knowledge utilization have not advanced understanding of the use of evidence in the policy process much beyond the decades-old National Research Council (1978) report. The family of suggestive concepts, typologies, and frameworks has yet to show with any reasonable certainty what changes have occurred in the nature, scope, and magnitude of the use of science as a result of different communication strategies or different forms of researcher-user collaborations (Dobbins et al., 2009b; Mitton et al., 2007). There is little assessment of whether innovations said to increase the use of science in policy have had or are having their desired effects.

A recent study reporting the results of a collaborative procedure among 52 participants covering a range of experiences in both science and policy identified 40 (!) key *unanswered* questions on the relationship between science and policy—this despite nearly four decades of research on the question of "use" (Sutherland et al., 2012). One extensive review of the literature reaches the striking conclusion that knowledge use is "so deeply embedded in organizational, policy, and institutional contexts that externally valid evidence pertaining to the efficacy of specific knowledge exchange strategies *is unlikely to be forthcoming*" (Contandriopoulos et al., 2010, p. 468 [italics added]).

Our conclusion is not that pessimistic. If "use" is broadly understood to mean that science—or, more specifically, in the language of evidence-based policy and practice, scientific evidence of the effectiveness of interventions—is incorporated into policy arguments, we agree that there probably will never be a definitive explanation of what strategies best facilitate or ensure that incorporation. But this conclusion does not rule out that the possibility that new approaches in the study of the science-policy nexus might reveal factors or conditions that have thus far been missed. Perhaps the preoccupation with defining use, identifying factors that influence it, and determining how to increase it has detracted from the search for alternative ways in which social science can contribute to understanding the use of science in policy. That possibility is the subject of Chapter 4.

4

Research on the Use of Science in Policy: A Framework

The use of science in policy is a human activity embedded in social processes and structures, a point now emphasized several times. We have emphasized as well that every field of science produces usable knowledge but explaining whether, how, and why that knowledge is used is a task of social science. This task leads us to ask what it means for science "to be of use" in policy. The relevant research literature on that question, summarized in Chapter 3, makes two central points:

1. Scientists are concerned with "improving use" by intensifying and strengthening research, specifically by developing stronger evidence of the effectiveness of social and technical interventions.
2. A scientific specialty on knowledge utilization is concerned with understanding precisely what "use" means and determining the relative weight of factors—timeliness, relevance, clarity and brevity of presentation, etc.—said to "increase the use" of science. It focuses on mechanisms for bridging the acknowledged gap between scientists and policy makers.

Both efforts have made major contributions to what we know about use. But we conclude that the inevitable indeterminacy and context-specific nature of use prevents these two efforts from providing a fully satisfactory understanding of the use of science or a satisfactory guide on how to strengthen that use in policy making.

This chapter provides a research agenda that, if seriously pursued, holds promise of providing a more satisfactory explanation and guide. We take our cue from an observation made 35 years ago by a deeply informed scholar (Weiss, 1978, p. 26):

> Social scientists tend to start out with the question: how can we increase the use of research in decision-making? They assume that greater use leads to improvement in decision-making. Decision makers might phrase it differently: how can we make wiser decisions, and to what extent, in what ways, and under what conditions, can social research help?

Weiss's own answer to her question frames the issue in a way the committee finds helpful (Weiss, 1978, p. 78):

> [H]ow to increase the use of social research in policy making is only one way to conceptualize the problem. An alternative view is: how can public policy making be improved, and what role can the social sciences play in that improvement? It may be that we have been concentrating too hard on the first formulation and not hard enough on the second.

Our proposed research framework is based on a view of policy makers engaged in an interactive, social process that assembles, interprets, and argues over science and whether it is relevant to the policy choice at hand and, if so, using that science as evidence supporting their policy arguments. Policy argument as a form of situated, practical reasoning directly leads to a concern with how evidence, in the specific way now defined, is *used* rather than how it is *produced.*

The research framework is presented under three headings: policy argumentation, psychological processes, and a systems perspective. Understanding science as evidence deployed in policy argument requires (1) investigating what makes good arguments in the policy domain— arguments that are accepted by policy makers as valid and sound—and the psychological processes influencing that acceptance; (2) investigating cognitive operations—mental models, schemata, prior knowledge, situated cognition, and related organizational circumstances—as well as institutional logics, practices, cultural assumptions (Coburn et al., in press; Hutchinson and Huberman, 1993; Spillane et al., 2002); and (3) investigating policy making from a systems perspective.

POLICY ARGUMENTATION

Policies result from practical arguments that offer reasons for taking a specific policy action (Ball, 1995). These practical deliberations (also referred to as policy arguments; see, e.g., Dunn, 1990; Fischer, 1980; 2007; Manzer, 1984; Marston and Watts, 2003; Stone, 2001) often involve what science says about likely outcomes of different policy choices. As emphasized in Chapter 1, they also involve political considerations insofar as policy choices influence who has and retains power and normative considerations regarding the desirability (or undesirability) of a proposed action, value judgments, and considerations of legitimacy (Esterling, 2004; Gasper, 1996).

Policy arguments have identifiable characteristics. For example, they are based on "a process through which diverse assumptions, interpretations, and contentions are commonly deliberated through an extended critical debate about policy recommendations and other proposals for public action" (Dunn, 1990, p. 324). Policy arguments generally constitute a package of considerations backed by reasons presented to persuade particular audiences of the validity of and need for a given action (Majone, 1989). The arguments consider not just the policy choice at hand, but how that policy interacts over time with many other policies—does opening a charter school in the community decrease or increase housing prices; do housing prices affect the local labor supply; does the labor supply affect whether a chain store locates in the community?

Obviously, it is a complex undertaking to sort out how the multiple characteristics of policy argument function together to yield a coherent, valid, and persuasive argument (Gasper, 1996; Hambrick, 1974; Toulmin, 1969). Although such an appraisal of policy arguments is necessary to understanding how science is used, that exercise is outside the scope of our report. It serves our purposes simply to emphasize that scientific findings, warrants, inferences, data and qualifications attached to these features of science are assembled in policy arguments in more or less compelling, fair, and balanced ways. This raises familiar issues: is relevant science ignored; does the quality and strength of evidence support the policy claims made; is evidence (pro and con) fully presented, etc.

More specifically, understanding how science is used in policy requires investigating what makes for reliable, valid, and compelling policy arguments *from the perspective of policy makers and those they need to persuade.* For example, arguments that certain consequences will follow from an intervention in a specific circumstance may involve a chain of reasoning with multiple premises. Surfacing and examining those premises and the extent to which they are accepted is critical to understanding whether the argument is perceived as valid (Cartwright, 2011). For arguments that involve statistical or probabilistic reasoning, it is critical to understand how probabilities are perceived and interpreted (Kahneman, 2011). It is necessary to investigate the ways in which argumentative strategies can mislead by making unwarranted assumptions, relying on unwarranted premises, or relying on fallacies in reasoning (Thouless, 1990; Toulmin, 1979) and, in general, why flawed arguments can nonetheless be persuasive.

We can now more explicitly see that science—data, findings, theories, concepts, and so on—becomes *evidence* when it is used in a policy argument. Although the term "evidence" so used is frequently encountered as claims about predicted or actual consequences—effects, impacts, outcomes or costs—of a specific action, that is but part of the story. Science can be used as evidence for early warning of a problem to be addressed (species loss, cyberterrorism, racial tensions), for target setting (gender pay equity, reduced school dropout rates), for implementation assessment (is it working here as it worked there), and for evaluation (cost-effective, unexpected outcomes).

It should now be clear that when use is the goal, focusing on producing good science is necessary but not sufficient. Strengthening the use of good science needs to take the next step of understanding how science is embedded in policy argumentation, and how science can provide the kind of information likely to inform these arguments. This directs attention to research in two areas: situated cognition (see, e.g., Anderson, Reder, and Simon, 1996; Elsbach, Barr, and Hargadon, 2005; Greeno, 1998; Spillane et al., 2002) and learning organizations (see, e.g., Moynihan and Landuyt, 2009; Senge, 1990).

Situated cognition is concerned with the interactions between cognitive schemata and organizational context—in which context (organizational rules, norms, resources, and procedures) is not simply a backdrop for the way users make sense of science as evidence, but actively influences and shapes cognitive processes, including creativity, innovation, learning, and strategic thinking. Situated cognition is a science relevant to organizational design supportive of continuous learning, critical thinking, and learning from experience and experimentation. Situated cognition emphasizes that learning is inseparable from doing, and thus is needed in examining the way researchers and stakeholders involved in addressing a particular problem collectively engage in learning about and solving that problem (Van

Langenhove, 2004). Social science can investigate situated cognition in organizations, as well as help policy-making organizations and groups operate as learning organizations (Common, 2004; Easterby-Smith, 2000; Gilson et al., 2009; Leeuw et al., 1994; Moynihan and Landuyt, 2009; Olsen and Peters, 1996; Vince and Broussine, 2000).

Keeping in mind that attention to policy argument is the necessary first step in constructing a research agenda relevant to understanding the use of science in policy, we turn to the second of our three components.

PSYCHOLOGICAL PROCESSES IN DECISION MAKING

There is an extensive literature in cognitive social psychology and behavioral decision theory on how people make judgments, decisions, and choices. Research is well developed in management sciences (e.g., Bazerman and Moore, 2008) and consumer behavior (e.g., Kivetz et al., 2008), and it has significant application in political science in the study of international relations and the making of foreign policy (e.g., Goldgeier and Tetlock, 2001; Jervis, 1976; Lau and Levy, 1998; Steinbruner, 1974).

These sciences have not, however, been applied to collective reasoning and group decision making *in public policy settings* at anything close to the level needed.[1] Of primary interest here are the branches of behavioral sciences that deal with social judgment theory (Cooksey, 1996), heuristics and biases (Kahneman, 2011; Tversky and Kahneman, 1974), learning and judgment making in teams (National Research Council, 2011b), and naturalistic decision making (Kahneman and Klein, 2009; Klein, 1998; Klein et al., 1993).

Research has deepened knowledge about the fallibility of human decision making, particularly the many cognitive biases to which people are subject (Kahneman, 2011). People have a proclivity to ignore evidence that contradicts their preconceived notions (confirmation bias); they may assess the frequency of an event by the ease with which instances are brought to mind (availability bias); and they may be overly cautious (loss aversion) (Kahneman et al., 2011; Tversky and Kahneman, 1974). Hypotheses about

---

[1]However, there is a research literature on group dynamics that deals with jury deliberations and other small group decision making, which includes sociological studies on such factors as peer pressure, perceived consensus, status differentiation, and gender differences. It constitutes a different theoretical and research tradition than the literature discussed here but also could be brought to bear on public policy decision making.

types of biases have been experimentally tested and extended by neuropsy-chology and evolutionary psychology (see, e.g., Gazzanaga, 2008).

How cognitive biases operate can be seen in an example from medical science. A medical practitioner explains why new research findings on the overuse and sometimes risky use of screenings for prostate cancer, colonos-copy for colon cancer, and mammogram testing will be ignored by most doctors (Bach, 2012, p. D5):

> Against the gravitational pull of doctor-knows-best culture . . . [g]uidelines written by academic types only impact the fringes of our practices. And despite the apparent move toward evidence-based medicine and comparative effectiveness research, most of us still feel that our own experiences and insights are the most relevant factors in medical decision-making.

Policy makers also inhabit a culture that stresses the importance of ex-perience and insight, and this culture is always at play when deciding how much to defer to "guidelines written by academic types." The social science that is needed to understand the use of science is not research about the consequences of those decisions: it is research about the decision process itself. This is true whether it is an individual decision maker, as in the medi-cal example, or, as is more often the case in policy decisions, a group-based decision.

A committee or agency making a policy decision may prematurely accept as true something that has been presented only as a possibility and then interpret existing data or seek out data confirming what has been de-cided (mindset or group-think biases). A dramatic example occurred among the scientists who advised President Gerald Ford on a swine flu vaccine (Neustadt and Fineberg, 1978). Research also shows "how close-knit groups can become so homogeneous that they do not realize limits to their in-group perspectives" (National Research Council, 2011a, p. 17), sometimes labeled the false consensus bias. Both individuals and groups mistakenly generalize to populations—say, people on welfare—on the basis of information readily accessible to them, such as the situation in their immediate neighborhood or anecdotes about "welfare queens."

Decision making in organizations is influenced by structures that aggregate and report information. These structures no less than indi-viduals can be biased. Institutionalized racism and sexism are well known examples. The 1986 Challenger space shuttle disaster was a consequence of organizational as well as technical deficiencies. That is, "the inability of various subunits in the National Aeronautics and Space Administration to integrate what each knew and from their different methods for processing information" (Zegart, 2011; cited in National Research Council, 2011a, p. 16). There are many ways that organizational factors "impair information integration," including "the need for secrecy, 'ownership' of information, everyday turf wars, intergroup rivalry, and differing skill sets. . . ." (National Research Council, 2011a, pp. 16-17).

Researchers who study cognitive biases do more than describe them. They study how biases can be overcome or circumvented (Kahneman, 2011; Kahneman et al., 2011). For example, the National Research Council (2011a, 2011b) has advised the Office of the Director of National Intelligence on how to improve intelligence assessments by recognizing group biases of intelligence analysts.

Bringing the insights of cognitive science to policy argument will present special challenges. In policy making, cognitive biases necessarily interact with values, norms, culture, and political power in ways unique to policy settings. Hammond (1996, pp. 264-265) describes the challenge in stark terms:

> the policy maker's task of integrating scientific information into the fabric of social values is an extraordinarily difficult task, for which there is no textbook, no handbook, no operating manual, no equipment, no set of heuristics, no theory, not even a tradition—unless a record of confusion can be called a tradition.

This challenge notwithstanding, behavioral decision theory and related fields can substantially increase understanding of policy argument and how science is used, misused, and ignored. Such understanding would be reason enough to recommend to cognitive scientists that they direct attention to "policy argumentation." But there is a further reason for including these fields in our research framework: it is becoming clear that cognitive science and behavioral economics can directly address policy design.

An example is the automatic contribution arrangements of the Pension Protection Act of 2006. This legislation, informed by behavioral economics, allows employers to enroll employees in a retirement savings plan (at a default contribution rate and default asset allocation) unless they explicitly opt out. This approach is in direct contrast to the previous arrangements in which employees were not enrolled unless they explicitly opted in. Introducing opt-out rules significantly increased employee participation in retirement savings plans (Beshears et al., 2010). For other examples of using knowledge about behavioral biases, see Congdon and Kling (2011), Orszag (2008), and Thaler and Sunstein (2008). Decision processes that increase stakeholders' commitments and public participation have also met with some success (see National Research Council, 2008).

The third component of our research agenda re-emphasizes that policy—and therefore the use of science in policy—unfolds in unusually complex settings. Greater emphasis must be placed on social science that takes this reality into account both in studying use and in researching solutions to social problems.

A SYSTEMS PERSPECTIVE

A report from the Institute of Medicine (2010a, pp. 5-6) noted:

> The real world is a complex system . . . many influences . . . are all interacting simultaneously. A systems perspective helps decision makers and researchers think broadly about the whole picture rather than merely studying the component parts in isolation. . . . A systems perspective can enhance the ability to develop and use evidence effectively and suggest actions with the potential to effect change. It can allow the forecasting of potential consequences of not taking action, possible unintended effects of interventions, the likely magnitude of the effect of one or more interventions, conflicts between or complementarity of interventions, and priorities among interventions.

A "systems perspective" is not one thing. It includes a number of approaches—complex systems, critical systems thinking, activity systems, and soft systems—and it includes various methodologies—agent-based modeling, microsimulation, systems dynamics modeling, and network analysis (see, e.g., Berry et al., 2002; Carrington et al., 2005; Christakis and Fowler, 2009; Epstein, 2006; Meadows, 2008; Miller and Page, 2007; Mitchell, 2009; Watts, 2003). The broad goal is "to provide insights into the way in which people, programs, and organizations interact with each other, their histories, and their environments" (Rogers and Williams, 2006, p. 80).

A number of policy areas have been studied from a systems perspective. For security policy, Jervis (1997) concludes that systems cannot be understood through examining only the attributes and goals of their elements. There are systems effects on individual actors and on the system as a whole, including emergent, indirect, and delayed effects, as well as unintended and unpredictable consequences from the interactivity of the system's elements. Concepts associated with studying complex systems—emergence, nonrecursive effects, adaptation—have been used to examine integration and innovation in primary health care organizations (North American Primary Care Research Group, 2009). A systems perspective has also been used to improve cooperative interaction in research communities and among researchers, policy makers, and public groups (see, e.g., Leischow et al., 2008; Midgley and Richardson, 2007). It has gained a strong foothold in evaluating complex social interventions (Eoyang and Berkas, 2007; Hargreaves, 2010; Midgley, 2007; Williams and Hummelbrunner, 2011). And it has been used in comparative cross-national studies of the use of science in regulatory policy making (Jasanoff, 2005). Menendian and Watt (2008) used concepts from systems theory to develop an understanding of contemporary racial conditions.

A recent white paper submitted to the National Science Foundation (Page, 2011) proposes that the social sciences develop methodologies for measuring and categorizing the complexity of social processes and structure interdisciplinary research to unpack how purposive actors respond to incentives, information, and cultural norms and how their psychological predispositions interact to produce social outcomes. The Office of Behavioral and Social Sciences Research at the National Institutes of Health (NIH) joined with 11 other NIH institutes in requesting research proposals to develop projects that use systems science methodologies relevant to understanding and explaining behavioral and social issues in health (described in Consortium of Social Science Associations, 2011). NIH also sponsored a minisymposium in July 2011 on how systems science can be used to inform public policy, using childhood obesity as an example.[2] More recently, NIH announced a funding opportunity to develop theory and methods to better understand complex social behavior through a systems perspective (National Institutes of Health, 2012b). Along the same lines is the call of the James S. McDonnell Foundation, as part of its 21st century science initiative, to develop tools for the study of complex, adaptive, nonlinear systems in a variety of fields, including biology, biodiversity, climate, demography,

---

[2]A videocast of the symposium, "Harnessing Systems Science Methodologies to Inform Pubic Policy: Systems Dynamics Modeling for Obesity Policy in the Envision Network," is available: http://videocast.nih.gov/summary.asp?file=16756 [February 2012].

epidemiology, technological change, economic development, governance, and computation. The 2008 Global Science Forum of the OECD focused on complexity science for public policy (OECD, 2009).

"Perhaps the most important location" where systems thinking is called for "is in making decisions and crafting policies that help navigate the complex structures that populate the world in which we live" (Sterman, 2006, p. 513). Moreover, because there is a lack of "a meaningful systems thinking capability," policies "often fail or worsen the problems they are intended to solve." In a world that is interconnected, "Systems thinking is an iterative learning process in which we replace a reductionist, narrow, short-run, static view of the world with a holistic, broad, long-term dynamic view, reinventing our policies and institutions accordingly" (Sterman, 2006, p. 509). A systems perspective is compatible with many forms of scientific investigation, including the effort to produce knowledge about the efficacy and effectiveness of policy interventions. Moreover, particular methods, such as agent-based modeling, can be evaluated with experimental designs to determine whether the interventions operate as expected. The proponents of system-based approaches recognize that experiments needed for these evaluations may be quite complex and that data may be based on simulations rather than measurement, but they have concluded that studies of complex systems should be anchored in sound quantitative methods.

Systems thinking is often important to understand the consequences of policies. A former assistant director of the National Science Foundation (Bradburn, 2004, p. 39) wrote:

> Governmental policies are blunt instruments to bring about social change. They almost never consider the dynamics put in motion by those changes. Thus, they inevitably suffer from unintended consequences. These unintended consequences are often large enough to nullify the positive effects of the policies or, even, to produce the opposite effect from that intended. . . . I approach [this issue] from the perspective of a social systems theorist and fault applications of social science analysis and research that fail to think through the dynamics of social systems and to pursue research that enables us to model more completely the effects of policy changes. I do not underestimate the difficulty of this task, but it is the direction that I think social sciences must be going.

CONCLUSION

We obviously strongly endorse social science continuing to improve its capacity to assess conditions and to help design and evaluate policies directed at those conditions. But this indispensible work provides little information about whether what is learned is used. Improving the scientific understanding of what occurs at the science-policy intersection involves going beyond the focus on what research "use" means and going beyond the effort to produce better science.

Social science has methods and theories that can significantly expand on whether what is learned is used, and can, in the process, add a new dimension to what science offers to policy. Our perspective urges broad social science attention to what happens during policy arguments, with a specific focus on whether, why, and how science is used as evidence in public policy.

5

The Next Generation of
Researchers and Practitioners

The three actors central to advancing and applying the research framework now outlined are (1) established scholars in the fields and specialties identified in Chapter 4; (2) Ph.D. candidates in those fields and specialties; and (3) administrators and faculty responsible for curricula in schools and programs summarized below by the term "policy education." For the first two of these actors, there are historical and contemporary models we briefly note; the third will involve fresh thinking.

ESTABLISHED SCHOLARS

In 1923, the Social Science Research Council (SSRC) was established to promote "co-operative research among the several disciplines" (Fosdick, 1952, p. 198),[1] as a necessary foundation for creating entirely new research fields and specialties. Later the term "field development" was coined. We use that term to describe a coordinated and well-funded effort to attract established scholars to important but under-researched issues for which their theories and methods are appropriate.

Citing metaphors favored in that more naïve time, the director of the Laura Spelman Rockefeller Memorial, Beardsly Ruml, who funneled millions of dollars to research universities and the SSRC in the 1920s, lamented that "All who work toward the general end of social welfare are embarrassed

---

[1]This early plea for interdisciplinary research did not use the term, which did not appear (as interdiscipline inquiries) until SSRC's *Sixth Annual Report*, 1929-1930 (noted by Sills, 1986).

by the lack of that knowledge which the social sciences must provide."
Ruml offered what for him was the clinching argument (cited in Fosdick,
1952, p. 194):

> It is as though engineers were at work without an adequate devel-
> opment of physics and chemistry, or as though physicians were
> practicing in the absence of the medical sciences. The direction of
> work in the social field is largely controlled by tradition, inspira-
> tion and expediency.

Ruml and the SSRC leadership had a clear goal: to professionalize the
social sciences, provide them methodological tools necessary for rigorous
research, and point them toward important fields of investigation.

Ruml was not naïve about the challenges: data were meager; research
was based on second-hand observations and anecdotal material; classroom
instruction isolated students from social conditions; and, especially, the social sciences were challenged to investigate topics that could not "be brought
into the laboratory for study," but "must be observed if, when, and as op-
erative." Difficulties notwithstanding, "unless means are found for meeting
the complex social problems that are so rapidly developing, our increasing
control of physical forces may be increasingly destructive of human values"
(cited in Fosdick, 1952, p. 195).

We bring this early philanthropic initiative to mind to draw a lesson
still applicable. Targeted funds can help develop new research specialties.
The well-funded SSRC emphasized interdisciplinary research and a strong
commitment to empirical methods. Social science researchers responded
not only to the SSRC, but also to the program priorities announced by other philanthropic foundations, to the Russell Sage Foundation in more than
a century of social science funding and to the larger foundations—Ford,
Carnegie, Hewlett, and MacArthur, among others—in the second half of
the 20th century. The label *field development,* for example, was attached
to area studies, a Cold War era success story. Coordinated conferences,
workshops, research monographs, and edited volumes advanced research
focused on showing how recently decolonized countries could engage in
"nation building" and how western democracies should meet the threat of
Communism, which in turn spawned a generation of research on the Soviet
Union and China watchers (so-called because lacking access to the Chinese
mainland, they "watched" from Hong Kong). There are many examples
of new research fields promoted by private foundations and government

funders—behavioral economics, human dimensions of climate change, population studies, life course development, aging, and race, ethnic, and gender studies.

An important example directly related to understanding the use of social science across a broad array of public policies is the ambitious effort pioneered by neoconservative social scientists who were skeptical about the effectiveness of many Great Society programs. With the Olin Foundation in the lead, private funds subsidized books, endowed university professorships, offered student fellowships, and established think tanks—similar to strategies earlier pioneered by the Spelman Rockefeller Memorial Fund—"all with the intent of changing the prevailing terms of debate" and advancing market-sympathetic policy (Rodgers, 2011, p. 7). This effort shaped the ongoing debates about the respective merits of the state and the market with respect to a long list of social policies—including whether poverty was better reduced by social welfare government programs or market forces, and whether school reform was better advanced through vouchers and school choice than leaving many educational practices under the influence of teacher unions.

Field development is not limited to foundations, though they have been particularly adept at it. The U.S. National Science Foundation (NSF) is attentive to how its funds might shape fields of inquiry, noting that in addition to its reliance on a well-established peer review process to guide its grant making, program officers should "identify promising research that responds to national priorities identified by Congress and the Administration" and to "incorporate agency or programmatic priorities" in NSF funding (Marrett, 2011, p. 3).[2] Particularly important to our purpose, the portfolio of grants funded by the NSF is expected to achieve "special program objectives and initiatives" and to build "capacity in a new and promising research area" (Marrett, 2011, p. 5).

A highly visible and in some quarters sharply criticized foray is the recent Science of Science Policy Initiative (Fealing et al., 2011). The broad purpose is to develop an evidentiary base for policy decisions on investments in basic and applied scientific research. New federal programs associated with this purpose include the Science of Science and Innovation Policy at NSF and an interagency task force sponsored by the National Science and Technology Council. A virtual community of practice has been organized,

---

[2]For a technical description of NSF's merit review policy, see http://www.nsf.gov/bfa/dias/policy/meritreview/ [February 2012].

facilitated by the establishment of a website hosted by the Office of Science and Technology Policy in the Executive Office of the President.[3] We cite this example not to endorse it (for a critical analysis, see Feller, 2011) but to illustrate how federal funding is used to develop new fields of inquiry: in this example, it is an attempt to build a community of practice among researchers and between researchers and policy makers.

Two sponsors of our study, the William T. Grant Foundation and the Spencer Foundation, have specifically targeted funding to better understand the use of research in policy and practice with respect to children and youth (W.T. Grant) and data and information to improve education (Spencer). The W.T. Grant Foundation sponsors research on the acquisition, interpretation, and use of research evidence to develop "strong theory and empirical evidence on when, how, and under what conditions research is used." Its request for proposals notes that "[r]esearch acquisition, interpretation, and use occurs within a social ecology" and that the foundation seeks "to understand how organizational, social, economic, and political contexts matter" (William T. Grant Foundation, 2012).[4]

The Spencer Foundation's Evidence for the Classroom Project, part of its broader Data Use and Educational Improvement Initiative, sponsors research on the assumptions behind data-based educational reforms "by investigating whether, when, and how student performance data informs instruction in K-8 classrooms." The goal is "to learn more about how K-8 teachers use student performance data for instructional decisions and how organizational and individual factors affect that use." Included in this initiative is research on how organizations learn and improve (Spencer Foundation, 2012).

In a review of Spencer-funded research papers published in the *American Journal of Education,* Goren (2012) notes that the papers "call for a deeper and better understanding of data, their use, the conditions that are most conducive for using data well, how individuals and groups of practitioners make sense of the data before them, and the intended and unintended consequences of data use for school improvement" (p. 233). The summary conclusion laments "that our understanding of how data lead to improvement in education is tremendously underdeveloped" (p. 234).

These Grant and Spencer examples are consonant with the research agenda described in Chapter 4. As valuable as they are, however, they

---

[3]For details, see http://www.scienceofsciencepolicy.net [February 2012].

[4]For a description of this research program and early lessons from its funded research, see Tseng (2012) and the accompanying commentary.

touch on a small subset of the issues that need study in order to develop a deeper and wider understanding of the use of science in the policy context. The Grant Foundation initiative is limited to children and youth, and the Spencer Foundation initiative is limited to data use as a particular feature of educational practice. Similarly, the NSF example noted above is limited to science policy, and it is narrow in its selection of research methods.

If these initiatives are joined by sponsored research on how science is used as evidence in many other policy areas—international security, economic growth, renewable energy, transportation efficiency, agricultural productivity, etc.—and are targeted to methods and approaches described in Chapter 4, a new research field on the scale of area studies or behavioral economics would take shape. Of course, established scholars have already worked out their future research, and we cannot expect more than a small percent to shift their interests to the framework in Chapter 4 (though we welcome being proven wrong). Science funders have long accepted this reality, and have often focused on entry-level researchers as better candidates for launching new fields and specialties. With this in mind, we turn next to Ph.D. candidates.

PH.D. TRAINING: AN ENTRY POINT

A well-tested strategy for establishing new research fields provides incentives early in a person's research career, especially at the dissertation phase. The SSRC pioneered this approach in the 1920s, eventually offering hundreds of fellowships in the social sciences that produced leaders in the academically based departments and in the steadily growing array of policy institutions (Fosdick, 1952, pp. 230-231). Another major chapter in the history of philanthropic leadership was the substantial, decades-long funding of graduate training in languages and area studies by the Ford and Mellon Foundations, in service of enlightened foreign policy. The 1958 National Defense Education Act (Title VI) added federal funds to this effort.

In more recent decades, dissertation grants provided by the MacArthur Foundation added depth to international security education, successfully reorienting the field from a 1960s focus on a limited array of issues, primarily arms control, to a broader consideration of how international economics, global immigration, climate change, and other "nonsecurity" issues were, in fact, deeply implicated in how the nation should approach its security challenges in the 21st century. The predoctoral research training program in the neurosciences, sponsored by the National Institutes of Health (NIH), encouraged broad, early-stage training in the neurosciences. This program was targeted to basic and disease-related research of importance to the participating institutes.[5] A successful current effort is NSF's Integrative Graduate Education and Research Traineeship (IGERT) Program. Initiated in 1997, the program is intended "to establish new models for graduate education and training in a fertile environment for collaborative research that transcends traditional disciplinary boundaries" (National Science Foundation, 2012c).

These funding initiatives, and there are others, have in common a determination to establish new fields by starting with researchers in the earliest stages of their professional training. The strategy rests on a simple assumption. Ph.D. candidates searching for a dissertation topic are attracted to new areas, where a single study can quickly be influential. The dissertation is the basis for their early publications, which, if cited, keeps them on this track. Enough young scholars on a similar track begin to establish a new field. This time-tested strategy fits with a central point of this report: attracting a fresh generation of researchers to studies of the use of science in policy should not be difficult in this period of heightened political (and, we expect, funder) attention to whether the substantial public investment in science—social sciences included—results in science that is used. The list of research topics is long—this is a small sample:

- Challenges in linking the natural and social sciences in the policy context;
- How variability in the quality of scientific evidence affects its use;
- The role of intermediaries in promoting evidence use;
- The responsiveness of policy makers to commissioned research;
- The interaction of scientific claims and value claims in policy argument; and
- Comparative research that considers how different government systems produce and use scientific evidence for policy and how this relates to differing political systems and beliefs about the role of government.

Based on their disciplinary training—in systems analysis, studies of complex organizations, science and technology studies, social psychology,

_____

[5]For details, see http://grants.nih.gov/grants/guide/pa-files/PAR-00-037.html [February 2012].

behavioral economics, political science, statistics, cognitive sciences, and the history of science—Ph.D. candidates can start with the substantial research literature on how scientific knowledge is *produced* and proceed quickly to what is *not* known about how science is *used* as evidence in policy making, and then apply methods and theories, already available from their disciplinary training, best suited to remedying the gaps in knowledge. These beginning scholars are guaranteed two attentive audiences for their work. There is an influential audience of public and private science funders, government agencies, institutes, think tanks, lobbyists, and others with a stake in whether relevant scientific knowledge is brought to bear in policy. The second audience is faculty responsible for what is being taught to students en route to careers in the policy enterprise, to which we now turn.

POLICY EDUCATION: WHAT IS NEEDED

Training beyond the bachelor's degree is a minimum job requirement for almost all public policy positions. Perhaps mentoring and on-the-job learning worked in an earlier period, when policy challenges slowly made their way to the public agenda and arrived as fairly straightforward questions of whether X leads to Y. That world, if it ever really existed, is clearly not today's policy world. A nation dependent on policy analysts and policy makers who learn as they go is put at risk when policy challenges (as well as information, both helpful and unhelpful) arrive at bewildering speed, from unexpected directions, and in ever more complex forms. Professional preparation is the norm today, and university-based programs are where that preparation occurs.

Senior policy positions often require (or assume) Ph.D.-level training, but a significant number of positions in the policy enterprise recruit from programs leading to a master of public administration (M.P.A.), the degree traditionally offered in schools of public policy, though now more likely to be labeled master of public policy (M.P.P.). This relabeling reflects the shift from careers in the civil service to those in the policy enterprise. Related training takes place in other professional schools, especially law, business, public health, social work, and education. There are also programs focused on particular policy arenas, such as environmental policy, security policy, and urban policy. Some of these have become stand-alone master's degrees, an increasing practice in higher education (Radin, 2000). Although the United States leads the world in establishing public policy schools and programs, similar initiatives are now found on every continent and in steadily growing numbers. We use the generic term "policy education" to cover the M.P.A., M.P.P., topical master's degrees, and related certificate programs.

This array of programs presents an obvious entry point for introducing fresh ways of thinking among those who will practice policy analysis and program design. Their education should be based on two priorities. One is now being taught—acquiring the competencies relevant to assessing policy-relevant research knowledge. One is not—developing a clear understanding of the factors that influence the conditions under which that knowledge is likely to be used.

These joint priorities distinguish policy education from what is provided in academic departments, where the priority is primarily the discovery of new knowledge—even recognizing that academic social scientists increasingly hope that their research will be used. Policy education also differs from what aspiring political consultants and policy advocates seek (though many looking for such careers earn a M.P.A. or M.P.P.), which are skills relevant to advancing a political cause or winning a policy battle. The academic social sciences adequately attend to the education of advanced students whose vocation is the discovery and dissemination of knowledge. The political world adequately supplies on-the-job training for those whose vocation is winning through bargaining and compromising, media campaigns, mobilization of support, and using science evidence selectively and tactically. Neither the academically oriented nor the politically motivated student is the audience we have in mind. Rather, it is the student whose priority is bringing scientific evidence to bear on policy choices, and wanting this not for tactical reasons but because it is a core professional principle. As Majone (1989, p. 7) writes:

> The job of analysts consists in large part of producing evidence and arguments to be used in the course of public debate. Its crucial argumentative aspect is what distinguishes policy analysis from academic social science on the one hand, and from problem-solving methodologies such as operations research on the other. The arguments that analysts produce may be more or less technical, more or less sophisticated, but they must persuade if they are to be taken seriously in the forums of public deliberation.

The statement of task guiding this report did not direct the committee to conduct a comprehensive investigation of what is being taught in policy programs and schools. Deliberations of the committee, however, led to the firm belief that it is timely to examine policy education in the same spirit that the famed Flexner Report (Flexner, 1910) examined medical education a century ago and the Ford Foundation (Gordon and Howell, 1959) and the Carnegie Corporation of New York (Pierson, 1959) examined business education in the 1950s. The Flexner Report, commissioned in 1908, stands out in this list; it is widely credited with initiating reforms that professionalized medical and health training appropriate for 20th century challenges, and from which the nation continues to benefit.

An analogous effort directed to policy education could determine if schools and programs are suitably aligned with the challenges that have emerged over the past half-century: decolonization; democratization; globalization; mass communication and the emergence of the Internet; economic and technological development; the international diffusion of science and technology; the rise of knowledge elites; and the growing influence of the private sector in information production and knowledge management, in addition to the host of specific competencies associated with evidence-based policy, performance metrics, cost-benefit analysis, and evaluation research. A Flexner-like effort could determine whether policy schools are providing the knowledge and skills relevant to assuring that policies responding to these broad challenges are influenced by science.

In the absence of such a study, we turn to a research literature offering partial though important insights into policy school objectives and the implementation of those objectives. In addition, the committee conducted its own cursory examination of the curricula of nearly 100 policy schools and programs in the United States. We acknowledge that what is readily available allows only best-guess estimates about what is being taught every year to the thousands of students enrolled in public policy courses. Although we would prefer to have a Flexner-like exhaustive study at hand, our immediate question can be adequately answered with what is available: how much of what we endorse as a policy education curriculum is already in place?

We are confident that practically all public policy education includes courses on the "politics of policy making." These courses draw on a large political science literature that examines how political considerations affect policy outcomes. There is also attention to the role of values, a topic appearing in any number of topical courses on the assumption that value tradeoffs appear in practically all policy choices. Examples include intergenerational choices, such as abundant energy for current generations versus the risk of sea-level rise that will inundate coastal communities of future generations; allocating public funds between competing public goods, such as repairing roads versus lower student-teacher ratios; deciding who should pay for policy failures, whether the costs of the collapse in the housing market should be borne by those who borrowed above their means or by those who packaged the mortgages in ways that hid the risks. More generally, students are taught that the complexity in policy making results not just from weighing counterarguments about effectiveness and efficiency, but also from facing questions about what is right, just, or fair.

If political and value considerations are being routinely taught, so are methods. In these courses there is a decided emphasis on quantitative skills. Morçöl and Ivanova (2010) document this, and categorize the quantitative methods courses into three groups: (1) research design courses, in which experimental and quasi-experimental designs are favored; (2) data collection methods, in which surveys are favored; and (3) analytic approaches, in which regression analysis is favored. That is, it is clear that methods associated with the "evidence-based policy" framework (see Chapter 3) are strongly represented in policy education curriculum. This is to be expected, and policy education should continue to emphasize the quantitative methods relevant to analyzing social conditions, designing responsive policy interventions, and evaluating the consequences of interventions.

However, as detailed in Chapter 4, other competencies are needed to navigate the policy world. These competencies include attention to the properties of reasoning about scientific knowledge (Grozer, 2009) and to understanding the assumptions underlying divergent policy framings, expert judgments, consensus-building techniques, and analytic methods or approaches. This knowledge will help prepare students to cope with the realistic, everyday problems encountered in applying existing knowledge—with its gaps, imperfections, and disciplinary constraints—to policy problems. Without such understanding, students may overestimate the persuasive power of scientific reasoning, and overlook the substantial barriers of institutional and cultural resistance to new research knowledge, unfamiliar policy framings, or solutions that challenge deeply held moral or ethical beliefs. Internships and case studies can help students learn about these and other complexities of the policy-making process.

Because the case study method is widely used in policy education, we reviewed a large number of case studies from the perspective of our report. Consistent with the observation above, cases used in policy schools routinely cover how political considerations influence policy outcomes and value tradeoffs. They draw student attention to the distribution of benefits and costs and how the "rules of the game" condition policy choices. They use to advantage a large number of key concepts and processes—from bu-reaucratic inertia to unintended consequences, from negotiation strategies to using the media.

What the examined cases rarely attend to is how scientific knowledge is used in policy making. There is little discussion of the quality or quantity of research available to the policy makers, even less discussion of whether that research is used as evidence, and still less about why science is ignored. Except incidentally, the cases do not explore the role of knowledge brokers or whether the ideas of evidence-based policy come into play. The pro-cesses and institutions through which policy makers gain access to relevant knowledge, such as expert advisory committees, receive little notice. There certainly is no attention to whether variation in cognitive biases of policy makers or variation in cultures of decision making tell them what to expect when science enters the policy argument. In summary, practically nothing of what is emphasized in Chapter 4 as ways to better understand the use of science is reflected in the case studies we examined.

An additional suggestive finding comes from Great Britain, where the current government has established the Behavioral Insights Team, a small office led by a social psychologist. Thaler (2012) describes how this office used a randomized control trial to test behavioral theory on when people conform to social norms. The issue was tax compliance; the treatment was a letter to late payers stating that others in their community pay their taxes promptly. There was a sharp increase in compliance in the treatment group, and not in the control group, whose message made no mention of neigh-bor's behavior. British tax authorities estimate that the reinforcing message could generate extra annual revenue of £30 million ($46.5m) nationwide. We cite this small study because the government (Thaler, 2012, p. 4) "is suf-ficiently convinced of the value of these activities [of the Behavioral Insights Team] that it announced last week that behavioral science is to be included in the required curriculum for civil servants." Behavioral science had not been taught in Britain's civil service training but now will be.

Though it would take a Flexner-style investigation to offer a thorough account of what is today being taught to thousands of M.P.A. and M.P.P. students in U.S. universities, our cursory review points to what is absent. Our review found few courses that draw on social psychology and cognitive science to provide public policy students with an understanding of human decision-making processes—including biases, heuristics, and probabilistic errors—as they pertain to reasoning about policy. Nor did we find many courses in which an anthropological, sociological, or humanistic approach to policy making is used to help students make sense of the interconnectedness of actors and institutions and the frameworks that shape policy choices. Nor did we find policy education to be self-conscious about the issue one might expect it to be most attentive to: what do students need to understand about the use of scientific evidence in public policy?

The social sciences have the opportunity to influence the competencies and perspectives that today's students in master's-level policy programs carry with them into positions across the policy enterprise. We hope that this report will spur self-examination across policy schools. One outcome might be differentiation, with some programs providing ever more rigorous training in methods and theories that strengthen research about "what works" and other programs emphasizing rigorous training in methods and theories that strengthen understanding of the conditions needed to put that research to policy use. Such a division of labor would result in a broad array of perspectives and skills available to think tanks, legislative staffs, policy units in executive branches, and other settings in the policy enterprise—from local government to international agencies, in both the public and the private sector.

There is no better way to summarize this chapter than repeating a truism—effective public policy is dependent on a steady supply of well-prepared graduates prepared for public service and associated careers in the policy enterprise. Our report advocates a broad definition of well prepared, certainly to include technical competencies in evaluation research, program design, measurement, and the like—but to include as well an understanding of how science can be used to inform public policy.

A CONCLUDING THOUGHT

The committee writes this report mindful that the American public's willingness to invest in science education and research is not unlimited, and that the immediate times emphasize scrutiny of the investment. But these times are also witness to a steadily growing policy enterprise—a broad effort to make "better" policy through the application of science. We have not taken a position on "better" policy, but have certainly taken a position on the value of, to return to our title, Using Science as Evidence in Public Policy. Moreover, we have written that it is within the competency of and is therefore an obligation of the social sciences to advance our understanding of "using science."

References

Agency for Healthcare Research and Quality. (2012). *Translating Research into Practice (TRIP)-II.* Available: http://www.ahrq.gov/research/trip2fac.htm [July 2012].

Anderson, J.R., Reder, L.M., and Simon, H.A. (1996). Situated learning and education. *Educational Researcher, 25*(4), 5-11.

Avey, P.C., Desch, M.C., Maliniak, D., Long, J.D., Peterson, S., and Tierney, M.J. (2012). The beltway vs. the ivory tower: Why academics and policy makers don't get along. *Foreign Policy,* January/February. Available: http://www.foreignpolicy.com/articles/2012/01/03/the_beltway_vs_the_ivory_tower [July 2012].

Bach, P.S. (2012). The trouble with "Doctor Knows Best." Health Section, *The New York Times,* June 5, p. D5.

Ball, W.J. (1995). A pragmatic framework for the evaluation of policy arguments. *Policy Studies Review, 14*(1/2), 1-24.

Baumgartner, F., and Jones, B.D. (1993). *Agendas and Instability in American Politics.* Chicago: University of Chicago Press.

Bazerman, M.H., and Moore, D.A. (2008). *Judgment in Managerial Decision Making.* New York: Wiley.

Berry, B.J.L., Kiel, L.D., and Elliott, E. (2002). Adaptive agents, intelligence, and emergent human organization: Capturing complexity through agent-based modeling. *Proceedings of the National Academy of Sciences, 99*(Supp. 3), 7,187-7,188.

Beshears, J., Choi, J., Laibson, D., Madrian, B.C., and Weller, B. (2010). Public policy and saving for retirement: The autosave features of the Pension Protection Act of 2006. In J.J. Siegfried (Ed.), *Better Living Through Economics* (pp. 274-288). Cambridge, MA: Harvard University Press.

Bogenschneider, K., and Corbett, T.J. (2010). *Evidence-Based Policymaking: Insights from Policy-Minded Researchers and Research-Minded Policymakers.* New York: Routledge.

Bourdieu, P. (1991). *Language and Symbolic Power.* Cambridge, MA: Harvard University Press.

Bradburn, N. (2004). What can public policies do? In OECD Directorate for Science, Technology, and Industry, *Re-Inventing the Social Sciences*. Paris: OECD. Available: http://www.oecd.org/LongAbstract/0,3425,en_2649_34269_33695705_1_1_1_1,00.html [July 2012].

Braybrooke, D., and Lindblom, C. (1963). *A Strategy of Decision*. New York: The Free Press.

Brewer, G.D., and deLeon, P. (1983). *The Foundation of Policy Analysis*. Chicago: Dorsey.

Bryk, A.S., Gomez, L.M., and Grunow, A. (2011). Getting ideas into action: Building networked improvement communities in education. In M.T. Hallinan (Ed.), *Frontiers in Sociology of Education* (pp. 127-162). New York: Springer.

Campbell, D.T. (1975). The social scientist as methodological servant of the experimenting society. In S.S. Nagel (Ed.), *Policy Studies and the Social Sciences* (pp. 27-32). New Brunswick, NJ: Transaction.

Caplan, N. (1979). The two communities theory and knowledge utilization. *American Behavioral Scientist, 22*(3), 459-470.

Carr, P., Dogan, E., Tirre, W., and Walton, E. (2007). Large-scale indicator assessments: What every educational policymaker should know. In P.A. Moss (Ed.), *Evidence and Decision Making, 106th Yearbook of the National Society for the Study of Education* (Part I, pp. 321-339). Malden, MA: Blackwell.

Carrington, P., Scott, J., and Wasserman, S. (Eds.). (2005). *Models and Methods in Social Network Analysis*. New York: Cambridge University Press.

Cartwright, N.D. (2011). Predicting what will happen when we act: What counts as warrant? *Preventive Medicine, 53*(4), 221-224.

Center for Responsive Politics. (2011). *Lobbying Database*. Available: http://www.opensecrets.org/lobby/ [August 2012].

Center on the Developing Child at Harvard University. (2007). *A Science-Based Framework for Early Childhood Policy: Using Evidence to Improve Outcomes in Learning, Behavior, and Health for Vulnerable Children*. Available: http://www.developingchild.harvard.edu [July 2012].

Chelimsky, E. (1991) Politics, policy making, data, and the homeless. *Housing Policy Debate, 2*(3), 683-697.

Christakis, N.A., and Fowler, J.H. (2009). *Connected: The Surprising Power of Our Social Networks and How They Shape Our Lives*. New York: Little, Brown and Company.

Coburn, C.E., Honig, M.I., and Stein, M.K. (in press). What is the evidence on districts' use of evidence? In J. Bransford, L. Gomez, D. Lam, and N. Vye (Eds.), *Research and Practice: Towards a Reconciliation*. Cambridge, MA: Harvard Educational Press.

Cochrane, A.L. (1972). *Effectiveness and Efficiency: Random Reflections of Health Services*. London, UK: Nuffield Provincial Hospitals Trust.

Cohen, D.M., and Garet, M. (1975). Reforming educational policy with applied social research. *Harvard Educational Review, 45*, 17-43.

Cohen, K., and Lindblom, C. (1979) *Usable Knowledge: Social Science and Social Problem Solving*. New Haven, CT: Yale University Press.

Coleman, J.S. (1966). *Equality of Educational Opportunity*. Washington, DC: U.S. Department of Health, Education and Welfare, Office of Education.

Common, R. (2004). Organisational learning in a political environment. *Policy Studies Journal, 25*(1), 35-49.

Congdon, W.J., and Kling, J.R. (2011). *Policy and Choice Public Finance Through the Lens of Behavioral Economics.* Washington, DC: Brookings Institution Press.

Conklin, A., Hallsworth, H., Hatziandreu, E., and Grant, J. (2008). *Briefing on Linkage and Exchange: Facilitating Diffusion of Innovation in Health Services.* Santa Monica, CA: RAND.

Consortium of Social Science Associations. (2011). Systems science and health in the behavioral and social sciences: Applications wanted. COSSA *Washington Update, 30*(16). Available: http://archive.constantcontact.com/fs021/1102766514430/archive/1107597363026.html [August 2012].

Contandriopoulos, D., Lemire, M., Denis, J-L., and Tremblay, E. (2010). Knowledge exchange processes in organizations and policy arenas: A narrative systematic review of the literature. *The Milbank Quarterly, 88*(4), 444-483.

Cooksey, R.W. (1996). *Judgment Analysis: Theory, Methods, and Applications.* San Diego: Academic Press.

Davies, H.T.O., Nutley, S.M., and Smith, P.C. (2000). *What Works: Evidence-based Policy and Practice in Public Services.* Bristol, UK: The Policy Press.

Dehue, T. (2001). Establishing the experimenting society: The historical origin of social experimentation according to the randomized controlled design. *American Journal of Psychology, 114*(2), 283-302.

DeLeon, P. (1988). *Advice and Consent: The Development of the Policy Sciences.* New York: Russell Sage.

Dobbins, M., Robeson, P., Ciliska, D., Hanna, S., Cameron, R., O'Mara, L., DeCorby, K., and Mercer, S. (2009a). A description of a knowledge broker role implemented as part of a randomized controlled trial evaluating three knowledge translation strategies. *Implementation Science, 4,* 23. Available: http://www.implementationscience.com/content/4/1/23 [August 2012].

Dobbins, M., Hanna, S.E., Ciliska, D., Manske, S., Cameron, R., Mercer, S.L., O'Mara, L., DeCorby, K., and Robeson, P. (2009b). A randomized controlled trial evaluating the impact of knowledge translation and exchange strategies. *Implementation Science, 4,* 61. Available: http://www.implementationscience.com/content/4/1/61 [August 2012].

Drolet, B.C., and Lorenzi, N.M. (2011). Translational research: understanding the continuum from bench to bedside. *Translational Research, 157*(1), 1-5.

Dryzek, J.S., and Bobrow, D.B. (1987). *Policy Analysis by Design.* Pittsburgh, PA: University of Pittsburgh Press.

Dunn, W.N. (1990). Justifying policy arguments: Criteria for practical discourse. *Evaluation and Program Planning, 13*(3), 321-329.

Dunworth, T., Hannaway, J., Holanhan, J., and Turner, M.A. (2008). *The Case for Evidence-Based Policy: Beyond Ideology, Politics, and Guesswork* (revised edition). Washington, DC: The Urban Institute. Available: http://www.urban.org/url.cfm?ID=901189 [July 2012].

Easterby-Smith, M. (2000). Organizational learning: Debates past, present and future. *Journal of Management Studies, 37*(6), 783-796.

Elsbach, K.D., Barr, P.S., and Hargadon, A.B. (2005). Identifying situated cognition in organizations. *Organization Science, 16*(4), 422-433.

Eoyang, G.H., and Berkas, T.H. (2007). Evaluation in a complex adaptive system. In B. Williams and I. Inman (Eds,), *Systems Concepts in Evaluation: An Expert Anthology.* Point Reyes Station, CA: Edge Press.

Epstein, J. (2006). *Generative Social Science: Studies in Agent-Based Computational Modeling.* Princeton: Princeton University Press.

Esterling, K.M. (2004). *The Political Economy of Expertise: Information and Efficiency in American National Politics.* Ann Arbor: University of Michigan Press.

Fealing, K.H., Lane, J.I., Marburger, J.H., and Shipp, S.S. (Eds.) (2011). *The Science of Science Policy: A Handbook.* Stanford, CA: Stanford University Press.

Feller, I. (2011). Science of science and innovation policy: The emerging community of practice. In K.H. Fealing, J.I. Lane, J.H. Marburger, and S.S. Shipp (Eds.), *The Science of Science Policy: A Handbook* (Ch. 8). Stanford, CA: Stanford University Press.

Finkelstein, A., Taubman, S., Wright, B., Bernstein, M., Gruber, J., Newhouse, J.P., Allen, H., Baicker, K., and Oregon Health Study Group (2012). The Oregon health insurance experiment: Evidence from the first year. *The Quarterly Journal of Economics, 127*(3), 1,057-1,106.

Fischer, F. (1980). *Politics, Values and Public Policy.* Boulder, CO: Westview Press.

Fischer, F. (2007). Deliberative policy analysis as practical reason: Integrating empirical and normative arguments. In F. Fischer, G. Miller, and M.S. Sidney (Eds.), *Handbook of Public Policy Analysis: Theory, Politics, and Methods* (pp. 223-236). Boca Raton, FL: CRC Press.

Fischer, F., and Forester, J. (Eds.). (1993). *The Argumentative Turn in Policy Analysis and Planning.* Durham, NC: Duke University Press.

Flexner, A. (1910). *Medical Education in the United States and Canada: A report to the Carnegie Foundation for the Advancement of Teaching.* New York: The Carnegie Foundation for the Advancement of Teaching. Available: http://www.carnegiefoundation.org/sites/default/files/elibrary/Carnegie_Flexner_Report.pdf [July 2012].

Floden, R.E., and Weiner, S.S. (1978). Rationality to ritual: The multiple roles of evaluation in governmental processes. *Policy Sciences, 9,* 9-18.

Fosdick, R.B. (1952). *The Story of the Rockefeller Foundation.* New York: Harper & Brothers.

Frantilla, A. (1998). *Social Science in the Public Interest: A Fiftieth-year History of the Institute for Social Research.* Issue 45 of the Bentley Historical Library Bulletin. Ann Arbor: University of Michigan, Bentley Historical Library.

Fuhrman, S. (1994). Uniting producers and consumers: Challenges in creating and utilizing educational research and development. In T.M. Tomlinson and A.C. Tuijnam (Eds.), *Educational Research and Reform: An International Perspective.* Washington, DC: U.S. Department of Education.

Gasper, D. (1996). Analyzing policy arguments. *European Journal of Development Research, 18*(1), 36-62.

Gazzanaga, M.S. (2008). *Human: The Science Behind What Makes Us Unique.* New York: HarperCollins.

Gigerenzer, G., Gaissmaier, W., Kurz-Milcke, E., Schwarz, L.M., and Woloshin, S. (2008). Helping doctors and patients make sense of health statistics. *Psychological Sciences in the Public Interest, 8*(2), 53-96.

Gilson, C., Dunleavy, P., and Tinkler, J. (2009). *Organizational Learning in Government Sector Organizations: Literature Review.* London, UK: LSE Policy Group.

Glasgow, R.E., and Emmons, K.M. (2007). How can we increase translation of research into practice? Types of evidence needed. *Annual Review of Public Health, 28*(4), 13-33.

Goldgeier, J.M., and Tetlock, P.E. (2001). Psychology and international relations theory. *Annual Review of Political Science, 4*, 67-92.

Gordon, R.A. and Howell, J.E. (1959). *Higher Education for Business.* Ford Foundation. New York: Columbia University Press.

Goren, P. (2012). Data, data, and more data—What's an educator to do? *American Journal of Education, 118*(2), 233-237.

Gormley, W.T., Jr. (2011). From science to policy in early childhood education. *Science, 333*(6,045), 978-981.

Green, L.A., and Seifert, C.M. (2005). Translation of research into practice: Why we can't "just do it." *Journal of the American Board of Family Medicine, 18*(6), 541-545.

Green, L.W. (2006). Public health asks of systems science: To advance our evidence-based practice, can you help us get more practice-based evidence. *American Journal of Public Health, 96*(3), 406-409.

Greenhalgh, T., Robert, G., Bate, P., Kyriakidou, O., Macfarlane, F., and Peacock, R. (2004). *How to Spread Good Ideas: A Systematic Review of the Literature on Diffusion, Dissemination, and Sustainability of Innovations in Health Service Delivery and Organisation.* London, UK: National Coordinating Centre for NHS Service Delivery and Organisation.

Greeno, J.G. (1998). The situativity of knowing, learning, and research. *American Psychologist, 53*(1), 5-26.

Grozer, T. (2009). Learning to reason about evidence and explanations: Promising directions in education. In E. Callan, T. Grozer, J. Kagan, R.E. Nisbett, D.N. Perkins, and L.S. Shulman (Eds.), *Education and Civil Society: Teaching Evidence-Based Decision Making* (pp. 51-74). Cambridge, MA: American Academy of Arts and Sciences.

Guston, D.H. (2000). *Between Politics and Science: Assuring the Integrity and Productivity of Research.* New York: Cambridge University Press.

Guttentag, M., and Struening, E.L. (Eds.). (1975). *Handbook of Evaluation Research, Vol. 1.* Beverly Hills, CA: Sage.

Hajer, M., and Wagenaar, H. (Eds.). (2003). *Deliberative Policy Analysis: Understanding Governance in the Networked Society.* Cambridge, UK: Cambridge University Press.

Hambrick, R.S. (1974). A guide for the analysis of policy arguments. *Policy Science, 5*(4), 469-478.

Hammond, K.R. (1996). *Human Judgment and Social Policy: Irreducible Uncertainty, Inevitable Error, Unavoidable Injustice.* New York: Oxford University Press.

Hargreaves, M. (2010). *Evaluating System Change: A Planning Guide.* Princeton, NJ: Mathematica Policy Research.

Heckman, J.J., and Vytlacil, E. (2005). Structural equations, treatment effects, and econometric policy evaluation. *Econometrics, 73*(3), 669-738.

Henig, J.R. (2009). Politicization of evidence: Lessons for an informed democracy. *Educational Policy, 23*(1), 137-160.

Henig, J.R. (in press). The politics of data use. *Teachers College Record, 114*(11), Special Issue, A. Bueschel and C. Coburn, Eds.

Hoppe. R. (1999). Policy analysis, science, and politics: From "speaking truth to power" to "making sense together." *Science and Public Policy, 26*(3), 201-210.

Huberman, M., and Cox, P. (1990). Evaluation utilization: Building links between action and reflection. *Studies in Educational Evaluation, 16*, 157-179.

Hutchinson, J., and Huberman, M. (1993, May). *Knowledge Dissemination and Utilization in Science and Mathematics Education: A Literature Review.* Arlington, VA: National Science Foundation.

Institute of Medicine. (2007). *The Learning Healthcare System. Workshop Summary.* L.A. Olsen, D. Aisner, and J.M. McGinnis, Eds. Roundtable on Evidence-Based Medicine. Washington, DC: The National Academies Press.

Institute of Medicine. (2010a). *Bridging the Evidence Gap in Obesity Prevention: A Framework to Inform Decision Making.* Committee on an Evidence Framework for Obesity Prevention Decision Making, S.K. Kumanyika, L. Parker, and L.J. Sim, Eds. Food and Nutrition Board. Washington, DC: The National Academies Press.

Institute of Medicine. (2010b). *Redesigning the Clinical Effectiveness Research Paradigm: Innovation and Practice-Based Approaches, Workshop Summary.* L. Olsen and J.M. McGinnis, Eds. Roundtable on Value & Science-Driven Health Care. Washington, DC: The National Academies Press.

Institute of Medicine (2011a). *Engineering a Learning Healthcare System: A Look at the Future, Workshop Summary.* C. Grossmann, W.A. Goolsby, L. Olsen, and J.M. McGinnis, Eds. Roundtable on Value & Science-Driven Health Care. Washington, DC: The National Academies Press.

Institute of Medicine (2011b). *Patients Charting the Course: Citizen Engagement in the Learning Health System, Workshop Summary.* L. Olsen, R.S. Saunders, and J.M. McGinnis, Eds. and Rapporteurs. Roundtable on Value & Science-Driven Health Care. Washington, DC: The National Academies Press.

Jasanoff, S. (1990). *The Fifth Branch: Science Advisors as Policymakers.* Cambridge, MA: Harvard University Press.

Jasanoff, S. (Ed.). (2004). *States of Knowledge: The Co-production of Science and Social Order.* London, UK: Routledge.

Jasanoff, S. (2005). *Designs on Nature: Science and Democracy in Europe and the United States.* Princeton, NJ: Princeton University Press.

Jasanoff, S., Markle, G.E., Peterson, J.C., and Pinch, T. (Eds.). (1995). *Handbook of Science and Technology Studies* (revised edition). Newbury Park, CA: Sage.

Jervis, R. (1976). *Perception and Misperception in International Politics.* Princeton, NJ. Princeton University Press.

Jervis, R. (1997). *System Effects: Complexity in Political and Social Life.* Princeton, NJ: Princeton University Press.

Johnson, K., Greenseid, L.O., Toal, S.A., King, J.A., Lawrenz, F., and Volkov, B. (2009). Research on evaluation use: A review of the empirical literature from 1986 to 2005. *American Journal of Evaluation, 30*(3), 377-410.

Kahneman, D. (2011). *Thinking, Fast and Slow.* New York: Farrar, Straus and Giroux.

Kahneman, D., and Klein, G. (2009). Conditions for intuitive expertise: A failure to disagree. *American Psychologist, 64*(6), 515-526.

Kahneman, D., Lovalo, D., and Sibony, O. (2011). The big idea: Before you make that big decision. *Harvard Business Review, 89*(6), 50-60.

Kingdon, J.W. (1984). *Agendas, Alternatives, and Public Policies.* Boston, MA: Little Brown.

Kivetz, R., Netzer, O., and Schrift, R. (2008). The synthesis of preference: Bridging behavioral decision research and marketing science. *Journal of Consumer Psychology, 18,* 179-186.

Klein, G. (1998). *Sources of Power: How People Make Decisions.* Cambridge, MA: MIT Press.

Klein, G., Oransanu, J., Calderwood, R., and Zsambok, C.E. (1993). *Decision Making in Action: Models and Methods.* Norwood, NJ: Ablex.

Landry, R., Amara, N., and Lamari, M. (2001). Utilization of social science research knowledge in Canada. *Research Policy, 30*(2), 333-349.

Lau, R.R., and Levy, J.S. (1998). Contributions of behavioural decision theory to political science. *Applied Psychology, 47*(1), 29-44.

Lavis, J.N. (2006). Research, public policymaking, and knowledge-translation processes: Canadian efforts to build bridges. *Journal of Continuing Education in the Health Professions, 26*(1), 37-45.

Lazer, D., Pentland, A., Adamic, L., Aral, S., Barabási, A-L., Brewer, D., Christakis, N., Contractor, N., Fowler, J., Gutmann, M., Jebara, T., King, G., Macy, M., Roy, D., and Van Alstyne, M. (2009). Computational social science. *Science, 323*(5,915), 721-723.

Leeuw, F.L., Rist, R.C., and Sonnichsen, R.C. (Eds.) (1994). *Can Governments Learn?: Comparative Perspectives on Evaluation and Organizational Learning.* New Brunswick, NJ: Transaction.

Leischow, S.J., Best, A., Trochim, W.M., Clark, P.I., Gallagher, R.S., Marcus, S.E., and Matthews, E. (2008). Systems thinking to improve the public's health. *American Journal of Preventive Medicine, 35*(2S), S196-S203.

Leviton, L.C., and Hughes, E.F.X. (1981). Research on the utilization of evaluations: A review and synthesis. *Evaluation Review, 5*, 524-548.

Lewis, K., and Burd-Sharps, S. (2010). *The Measure of America 2010-2011: Mapping Risks and Resilience.* New York: New York University Press and Social Science Research Council.

Lindblom, C. (1959). The science of muddling through. *Public Administration Review, 19*, 79-88.

Lindblom, G.E. (1968). *The Policy-Making Process.* Englewood Cliffs, NJ: Prentice Hall.

Louis, K.S., Blumenthal, D., Gluck, M.E., and Stoto, M.A. (1989). Entrepreneurs in academe: An exploration of behaviors among life scientists. *Administrative Science Quarterly, 34*(1), 110-131.

Lowrey, A. (2011). Programs that tie funds to effectiveness are at risk. *The New York Times,* p. A13, December 3. Available: http://www.nytimes.com/2011/12/03/us/politics/programs-tying-us-funds-to-effectiveness-are-at-risk.html?_r=0 [October 2012].

Lynd, R.S. (1939). *Knowledge for What? The Place of Social Science in American Culture.* Princeton, NJ: Princeton University Press.

Majone, G. (1989). *Evidence, Argument and Persuasion in the Policy Process.* New Haven, CT: Yale University Press.

Manzer, R. (1984). Public policymaking as practical reasoning. *Canadian Journal of Political Science, 17*(3), 577-594.

Marrett, C. (2011). *The Merit Review Process: Ensuring Limited Federal Resources Are Invested in the Best Science.* Testimony before the Subcommittee on Research and Science Education, Committee on Science, Space, Technology, U.S. House of Representatives, July 26. Available: http://science.house.gov/sites/republicans.science.house.gov/files/documents/hearings/072611_Marrett.pdf [July 2012].

Marston, G., and Watts, R. (2003). Tampering with the evidence: A critical appraisal of evidence-based policy making, *The Drawing Board: An Australian Review of Public Affairs, 3*(3), 143-163.

McDonnell, L.M. (2009). Repositioning politics in education's circle of knowledge. *Educational Researcher, 38*(6), 417-427.

McDonnell, L.M., and Weatherford, M.S. (2012). *Evidence Use and Stages of the Common Core State Standards Movement.* Paper prepared for presentation at the Annual Meeting of the American Educational Research Association, Vancouver, April 16.

McGann, J.G. (2010). *The Global "Go-To Think Tanks," 2009* (The Think Tanks and Civil Societies Program). Available: http://www.sas.upenn.edu/irp/documents/2009Global GoToReportThinkTankIndex1.31.10_2010.02.14.pdf [February 2011].

Meadows, D.H. (2008). *Thinking in Systems: A Primer.* Hartland, VT: Sustainability Institute.

Menendian, S., and Watt, C. (2008). *Systems Primer: Systems Thinking and Race.* The Kirwan Institute for the Study of Race and Ethnicity. Columbus, OH: Ohio State University.

Midgley, G. (2007). Systems thinking in evaluation. In B. Williams and I. Inman (Eds.), *Systems Concepts in Evaluation: An Expert Anthology.* Point Reyes Station, CA: Edge Press.

Midgley, G., and Richardson, K.A. (2007). Systems thinking for community involvement in policy analysis. *Emergence: Complexity and Organization, 9*(1-2), 167-183.

Miller, J.H., and Page, S.E. (2007). *Complex Adaptive Systems: An Introduction to Computational Models of Social Life.* Princeton, NJ: Princeton University Press.

Mitchell, M. (2009). *Complexity: A Guided Tour.* New York: Oxford University Press.

Mitton, C., Adair, C.E., McKenzie, E., Patten, S.B., and Perry, B.W. (2007) Knowledge transfer and exchange: Review and synthesis of the literature. *The Milbank Quarterly, 85*(4), 729-768.

Morçöl, G., and Ivanova, N.P. (2010). Methods taught in public policy programs: Are quantitative methods still prevalent? *Journal of Public Affairs Education, 16*(2), 255-277.

Mosteller, F., and Boruch, R.F. (2002). *Evidence Matters: Randomized Trials in Education Research.* Washington, DC: Brookings Institution Press.

Mosteller, F., and Moynihan, D.P. (Eds.). (1972). *On Equality of Educational Opportunity.* New York: Random House.

Moynihan, D.P., and Landuyt, N. (2009). How do public organizations learn? Bridging structural and cultural perspectives. *Public Administration Review, 69*(6), 1,097-1,105.

National Academy of Education. (1999). *Recommendations Regarding Research Priorities: An Advisory Report to the National Educational Research Policy and Priorities Board.* Washington, DC: National Academy of Education. Available: http://www.naeducation.org/Research_Priorities_Publication.pdf [July 2012].

National Institutes of Health. (2012a). *Budget Request, Office of Budget.* Available: http://officeofbudget.od.nih.gov/br.html# [July 2012].

National Institutes of Health. (2012b). *Modeling Social Behavior.* Funding Opportunity Announcement RFA-GM-13-006. Available: http://grants.nih.gov/grants/guide/rfa-files/RFA-GM-13-006.html [July 2012].

National Research Council. (1978). *Knowledge and Policy: The Uncertain Connection,* L. Lynn, Jr. (Ed.). Washington, DC: National Academy of Sciences.

National Research Council. (1999a). *Improving Student Learning: A Strategic Plan for Education Research and Its Utilization.* Committee on a Feasibility Study for a Strategic Education Research Program, Commission on Behavioral and Social Sciences and Education. Washington, DC: National Academy Press.

National Research Council. (1999b). *Sowing Seeds of Change: Informing Public Policy in the Economic Research Service of USDA.* Panel to Study the Research Program of the Economic Research Service, J.F. Geweke, J.T. Bonnen, A.A. White, and J.J. Koshel, Eds. Committee on National Statistics, Commission on Behavioral and Social Sciences and Education. Washington, DC: National Academy Press.

National Research Council. (2008). *Public Participation in Environmental Assessment and Decision Making.* Panel on Public Participation in Environmental Assessment and Decision Making, T. Dietz and P.C. Stern, Eds. Committee on the Human Dimensions of Global Change, Division of Behavioral and Social Sciences and Education. Washington, DC: The National Academies Press.

National Research Council. (2009). *Principles and Practices for a Federal Statistical Agency, Fourth Edition.* Committee on National Statistics. C.F. Citro, M.E. Martin, and M.L. Straf, Eds. Division of Behavioral and Social Sciences and Education. Washington, DC: The National Academies Press.

National Research Council. (2011a). *Intelligence Analysis for Tomorrow: Advances from the Behavioral and Social Sciences.* Committee on Behavioral and Social Science Research to Improve Intelligence Analysis for National Security, Board on Behavioral, Cognitive, and Sensory Sciences. Division of Behavioral and Social Sciences and Education. Washington, DC: The National Academies Press.

National Research Council. (2011b). *Intelligence Analysis for Tomorrow: Behavioral and Social Scientific Foundations.* Committee on Behavioral and Social Science Research to Improve Intelligence Analysis for National Security, B. Fischhoff and C. Chauvin, Eds. Board on Behavioral, Cognitive, and Sensory Sciences. Division of Behavioral and Social Sciences and Education. Washington, DC: The National Academies Press.

National Science Foundation. (2012a). *Federal Funds for Research and Development: Fiscal Years 2009-11.* [NSF 12-318]. National Center for Science and Engineering Statistics. Available: http://nsf.gov/statistics/nsf12318/pdf/tab19.pdf [July 2012].

National Science Foundation. (2012b). *FY 2013 Budget Request to Congress.* Social, Behavioral, and Economic Sciences. Available: http://www.nsf.gov/about/budget/fy2013/toc.jsp [July 2012].

National Science Foundation. (2012c). *Integrative Graduate Education and Research Traineeship Program (IGERT).* Available: http://www.nsf.gov/funding/pgm_summ.jsp?pims_id=12759 [July 2012].

Neilson, S. (2001). *IDRC-Supported Research and its Influence on Public Policy—Knowledge Utilization and Public Policy Processes: A Literature Review.* Ottawa, Canada: International Development Research Centre. Available: http://www.idrc.ca/uploads/user-S/11303580541IDRC_supported_research_and_influence_on_public_policy.pdf [July 2012].

Neustadt, R.E., and Fineberg, H.V. (1978). *The Swine Flu Affair: Decision Making on a Slippery Disease.* Washington, DC: National Academy Press.

North American Primary Care Research Group. (2009). *Learning About Complexity Science.* Available: http://www.napcrg.org/Beginner%20Complexity%20Science%20Module.pdf [August 2012].

Nutley, S., and Webb, J. (2000). Evidence and the policy process. In H.T.O. Davies, S.M. Nutley, and P.C. Smith (Eds.), *What Works?: Evidence-Based Policy and Practice in Public Services* (pp. 13-41). Bristol, UK: The Policy Press.

Nutley, S., Walter, I., and Bland, N. (2002). The institutional arrangements for connecting evidence and policy: The case of drug misuse. *Public Policy and Administration, 17*(3), 76-94.

Nutley, S., Walter, I. and Davies, H.T.O. (2007). *Using Evidence: How Research Can Inform Public Services.* Bristol, UK: The Policy Press.

O'Connor, A. (2007). *Social Science for What?* New York: Russell Sage Foundation.

OECD. (2009). *Applications of Complexity Science for Public Policy: New Tools for Finding Unanticipated Consequences and Unrealized Opportunities.* Paris: OECD.

Oh, C.H. (1997). Issues for the new thinking of knowledge utilization: Introductory remarks. *Knowledge, Technology, and Policy, 10*(3), 3-10.

Olsen, J.P., and Peters, B.G. (Eds.). (1996). *Lessons from Experience: Experiential Learning in Administrative Reforms in Eight Democracies.* Oslo, Norway: Scandinavian University Press.

Orszag, P. (2008). *Behavioral Economics: Lessons from Retirement Research for Health Care and Beyond.* Presentation to the Retirement Research Consortium. Available: http://www.cbo.gov/ftpdocs/96xx/doc9673/Presentation_RRC.1.1.shtml [February 2012].

Page, S.E. (2011). *Complexity in Social Political and Economic Systems.* SBE 2020 7 white paper, Directorate for Social, Behavioral and Economic Sciences, National Science Foundation. Available: http://www.nsf.gov/sbe/sbe_2020/submission_detail.cfm?upld_id=97 [August 2012].

Pelz, D.C., and Andrews, F.M. (1976). *Scientists in Organizations: Productive Climates for Research and Development.* New York: Wiley.

Pierson, F. (1959). *A Study of University-College Programs in Business Administration.* Carnegie Corporation of New York. New York: McGraw-Hill.

Radin, B. (2000). *Beyond Machiavelli: Policy Analysis Comes of Age.* Washington, DC: Georgetown University Press.

Rein, M., and White, S.H. (1977). Policy research: belief and doubt. *Policy Analysis, 3*(2), 239-271.

Renn, O. (1995). Style of using scientific expertise: A comparative framework. *Science and Public Policy, 22*(3), 147-156.

Rich, R.F. (1997). Measuring knowledge utilization: Processes and outcomes. *Knowledge, Technology, and Policy, 10*(3), 11-24

Riecken, H.W., and Boruch, R.F. (Eds.). (1975). *Social Experimentation.* New York: Academic Press.

Rodgers, D.T. (2011). *Age of Fracture.* Cambridge, MA: The Belknap Press of Harvard University Press.

Rogers, P.J., and Williams, B. (2006). Evaluation for practice improvement and organizational learning. In I.F. Shaw, J.C. Greene, and M.M. Mark (Eds.), *Handbook of Evaluation* (pp. 76-97). London, UK: Sage.

Rosenblatt, A., and Tseng, V. (2010) The demand side: Uses of research in child and adolescent mental health services. *Administration and Policy in Mental Health and Mental Health Services Research, 37,* 201-204.

Ruttan, V.W. (1984). Social science knowledge and institutional change, *American Journal of Agricultural Economics, 66*(55), 549-559.

Sabatier, P.A. (2007). *Theories of the Policy Process.* Boulder, CO: Westview Press.

Sabatier, P.A., and Jenkins-Smith, H.C. (Eds.). (1993). *Policy Change and Learning: An Advocacy Coalition Approach.* Boulder, CO: Westview Press.

Schumpeter, J.A. (1942). *Capitalism, Socialism and Democracy.* New York: Allen and Unwin.

Senge, P. (1990). *The Fifth Discipline: The Art and Practice of the Learning Organization.* New York: Doubleday.

Shulha, L.M. and Cousins, J.B. (1997), Evaluation use: Theory, research, and practice. *American Journal of Evaluation, 18*(3).

Sibley, E. (1974). *Social Science Research Council: The First Fifty Years.* New York: Social Science Research Council.

Sills, D.L. (1986). A note on the origin of interdisciplinary. *Items, 40*(1), 17-18.

Silver, H.J., Sharpe, A.L., Kelly, H., Kobor, P., and Sroufe, G.E. (2012). Social and behavioral science research in the FY 2013 budget. *AAAS REPORT XXXVII: Research and Development FY 2013* (Ch. 19, pp. 205-213). Report of the Intersociety Working Group. Washington, DC: American Association for the Advancement of Science. Available: http://www.aaas.org/spp/rd/rdreport2013/13pch19.pdf [July 2012].

Simon, H. (1957). *Models of Man: Social and Rational.* New York: Wiley.

Sims, C.A. (2010). But economics is not an experimental science. *Journal of Economic Perspectives, 24*(2), 59-68. Available: http://www.aeaweb.org/articles.php?doi=10.1257/jep.24.2.59 [July 2012].

Slavin, R.E. (2006). Translating research into widespread practice: The case of success for all. In M. Constas and R. Sternberg (Eds.), *Translating Theory and Research into Educational Practice* (pp. 113-126). Mahwah, NJ: Erlbaum.

Smith, M.S., and Smith, M.L. (2009) Research in the policy process. In G. Sykes, B. Schneider, and D.N. Plank (Eds.), *Handbook of Education Policy Research.* New York: Routledge.

Social Science Research Council. (2012). *Measure of America.* Available: http://www.measureofamerica.org/project/ [October 2012].

Spencer Foundation. (2012). *Strategic Initiatives: Data Use and Educational Improvement.* Available: http://www.spencer.org/content.cfm/data-use-and-educational-improvement [July 2012].

Spillane, J.P., and Miele, D.B. (2007). Evidence in practice: A framing of the terrain. In P.A. Moss (Ed.), *Evidence and Decision Making, 106th Yearbook of the National Society for the Study of Education* (Part I, pp. 46-73). Malden, MA: Blackwell.

Spillane, J.P., Reiser, B.J., and Reimer, T. (2002). Policy implementation and cognition: Reframing and refocusing implementation, *Review of Educational Research, 72*(3), 387-431.

State of the USA. (2012). *The State of the USA.* Available: http://www.stateoftheusa.org/ [October 2012].

Steinbruner, J. (1974). *The Cybernetic Theory of Decision.* Princeton, NJ: Princeton University Press.

Sterman, J.D. (2006). Learning from evidence in a complex world. *American Journal of Public Health, 96*(3), 505-514.

Stinchcombe, A. (1987). *The Logic of Scientific Inferences. Constructing Social Theories.* Chicago: University of Chicago Press.

Stokes, D.E. (1997). *Pasteur's Quadrant: Basic Science and Technological Innovation.* Washington, DC: Brookings Institution Press.

Stone, D. (2001). *The Policy Paradox: The Art of Political Decision Making* (revised edition). New York: W.W. Norton.

Stone, D., with Maxwell, S., and Keating, M. (2001). *Bridging Research and Policy, An International Workshop.* Funded by the UK Department for International Development, Warwick University. Warwick, UK: Centre for the Study of Globalisation and Regionalisation.

Sudsawad, P. (2007). *Knowledge Translation: Introduction to Models, Strategies, and Measures.* Austin, TX: Southwest Educational Development Laboratory, National Center for the Dissemination of Disability Research. Available: http://www.ncddr.org/kt/products/ktintro/ [July 2012].

Sutherland, W.J., Bellingan, L., Bellingham, J.R., Blackstock, J.J., Bloomfield, R.M., Bravo, M., Cadman, V.M., Cleevely, D.D., Clements, A., Cohen, A.S., Cope, D.R., Daemmrich, A.A, Devecchi, C., Anadon, L.D., Denegri1, S., Doubleday, R., Dusic, N.R., Evans, R.J., Feng, W.Y., Godfray, H.C.J., Harris, P., Hartley, S.E., Hester, A.J., Holmes, J., Hughes, A., Hulme, M., Irwin, C., Jennings, R.C., Kass, G.S., Littlejohns, P., Marteau, T.M., McKee, G., Millstone, E.P., Nuttall, W.J., Owens, S., Parker, M.M., Pearson, S., Petts, J., Ploszek, R., Pullin, A.S., Reid, G., Richards, K.S., Robinson, J.G., Shaxson, L., Sierra, L., Smith, B.G., Spiegelhalter, D.J., Stilgoe, J., Stirling, A., Tyler, C.P., Winickoff, D.E., and Zimmern, R.L. (2012). A collaboratively-derived science-policy research agenda. *PLoS ONE, 7*(3): e31824. Available: http://www.plosone.org/article/info%3Adoi%2F10.1371%2Fjournal.pone.0031824 [July 2012].

Szanton, P. (2001). *Not Well Advised: The City as Client—An Illuminating Analysis of Urban Governments and Their Consultants.* San Jose, CA: Authors Choice Press. (Orig. pub. New York: Russell Sage Foundation, 1981).

Thaler, R.H. (2012). Watching behavior before writing the rules. *The New York Times,* Sunday Money, July 8. Available: http://www.nytimes.com/2012/07/08/business/behavioral-science-can-help-guide-policy-economic-view.html?pagewanted=all [October 2012].

Thaler, R.H., and Sunstein, C.R. (2008). *Nudge: Improving Decisions about Health, Wealth, and Happiness.* New Haven CT: Yale University Press.

Thouless, R. (1990). *Straight and Crooked Thinking* (revised edition). London, UK: Hodder Arnold H&S.

Toulmin, S. (1969). *The Uses of Argument* (updated edition). Cambridge, UK: Cambridge University Press.

Toulmin, S. (1979). *An Introduction to Reasoning.* New York: Macmillan.

Tseng, V. (2012). The uses of research in policy and practice, *Social Policy Report, 26*(2), 3-16.

Tversky, A., and Kahneman, D. (1974). Judgment under uncertainty: Heuristics and biases. *Science, 185,* 1,124-1,131.

United Nations. (2012). *Human Development Index.* Available: http://hdr.undp.org/en/statistics/hdi/ [July 2012].

U.S. Department of Education. (2012). *The Regional Educational Laboratory Program (REL) About Us.* Available: http://ies.ed.gov/ncee/edlabs/about/ [July 2012].

U.S. Office of Management and Budget. (2011). *Statistical Programs of the United States Government: Fiscal Year 2012.* Washington, DC: U.S. Government Printing Office.

Van Langenhove, L. (2004). From opening to rethinking the social sciences. In *Re-inventing the Social Sciences*, Report of a workshop, Lisbon, November 8-9, 2001. OECD Directorate for Science, Technology, and Industry. Paris: OECD. Available: http://www.oecd.org/science/scienceandtechnologypolicy/33695704.pdf [October 2012].

Vince, R., and Broussine, M. (2000). Rethinking organisational learning in local government. *Local Government Studies, 26*(1), 15-30.

Watts, D.J. (2003). *Small Worlds: The Dynamics of Networks between Order and Randomness (Princeton Studies in Complexity)*. Princeton, NJ: Princeton University Press.

Webber, D.J. (1991). The distribution and use of policy knowledge in the policy process. *Knowledge and Policy, 4*(4), 6-35.

Weiss, C.H. (Ed.). (1977). *Using Social Research in Public Policy Making*. Lexington, MA: Lexington Books.

Weiss, C.H. (1978). Improving the linkage between social research and public policy. In L.E. Lynn (Ed.), *Knowledge and Policy: The Uncertain Connection*. Washington, DC: National Academy Press.

Weiss, C.H. (1979). The many meanings of research utilization. *Public Administration Review, 39*(5), 426-431.

Weiss, C.H. (1991). Policy research as advocacy: Pro and con. *Knowledge, Technology and Policy, 4*(1-2), 37-55.

Weiss, C.H. (1998). Have we learned anything new about the use of evaluation? *American Journal of Evaluation, 23*(2), 137-150.

Weiss, C.H., and Bucuvalas, M.J. (1980). *Social Science Research and Decision Making*. New York: Columbia University Press.

Weiss, C.H, Murphy-Graham, E., and Birkeland, S. (2005). An alternate route to policy influence: Evidence from a study of the Drug Abuse Resistance Education (D.A.R.E.) Program. *American Journal of Evaluation, 26*(1), 12-31.

William T. Grant Foundation. (2012). *Request for Proposals (RFP) on Understanding the Acquisition, Interpretation, and Use of Research Evidence in Policy and Practice*. Available: http://www.wtgrantfoundation.org/funding_opportunities/research_grants/rfp_for_the_use_of_research_evidence [August 2012].

Williams, B., and Hummelbrunner, R. (2011). *Systems Concepts in Action*. Stanford, CA: Stanford University Press.

Wilson, J.Q. (1996). Foreword. In M. Gerson (Ed.), *The Essential Neoconservative Reader*. Reading, MA: Addison-Wesley.

Wilson, W. (1901). *Congressional Government: A Study in American Politics*. Boston, MA: Houghton, Mifflin and Company.

Zegart, A.B. (2011). Implementing change: Organizational challenges. In National Research Council, *Intelligence Analysis for Tomorrow: Behavioral and Social Scientific Foundations* (pp. 309-329). B. Fischhoff and C. Chauvin (Eds.), Committee on Behavioral and Social Science Research to Improve Intelligence Analysis for National Security, Board on Behavioral, Cognitive, and Sensory Sciences. Division of Behavioral and Social Sciences and Education. Washington, DC: The National Academies Press.

Appendix A

Selected Major Social Science Research Methods: Overview

T he social sciences comprise a vast array of research methods, models, measures, concepts, and theories. This appendix provides a brief overview of five common research methods or approaches and their assets and liabilities: experiments, observational studies, evaluation, meta-analyses, and qualitative research. We close with a discussion of new sources of data. We begin with a brief comment on cause and effect.

To inform public policy, researchers often frame their studies in terms of causal conclusions and reason from an intervention to its intended outcomes. Many types of research methods are used for this purpose, as well as statistical analyses.

Research that can reach causal conclusions has to involve well-defined concepts, careful measurement, and data gathered in controlled settings. Only through the accumulation of information gathered in a systematic fashion can one hope to disentangle the aspects of cause and effect that are relevant to a policy setting. Statistical methodology alone is of limited value in the process of inferring causation.

The literature on causality spans philosophy, statistics, and social and other sciences. Our use here is consistent with the recent literature describing causality in terms of counterfactuals, interventions or manipulation, and probabilistic interpretations of causation.

EXPERIMENTS

In the simplest study of an intervention, one group of subjects who receive the intervention (the *treatment group*) is compared with another group of subjects (the *control group*) who do not. When the control group receives no other intervention, it serves to depict the *counterfactual*: what would happen in the absence of the intervention. Many studies, however, are more elaborate and may involve multiple interventions and controls.

An experiment is a study in which the investigator controls the selection of the subjects who may receive the intervention and assigns them to treatment and control groups at random. Experiments can be conducted in highly controlled settings, such as in a laboratory, or in the field, such as at a school, so as to better reflect the context in which an intervention would be implemented in practice. The former assess *efficacy*, or whether the intervention produces the intended effect. The latter, called *randomized controlled field trials* (RCFTs), assess *effectiveness*, or whether the intervention produces the intended effect in practice.

One important advantage of RCFTs is that secondary variables do not confound the effects of an intervention. That is, in an ideal study, an investigator wants to compare the effects of an intervention on a treatment group that is as similar as possible to the control group in all important respects except for having received the intervention. But this ideal can be affected by secondary or intervening variables—other factors by which the treatment group differs from the control group but are not of primary interest—which confound the effects of the intervention. These factors can influence the outcome of an experiment. In an RCFT, however, these secondary variables do not necessarily need to be controlled for in the design or the analysis: randomization obviates even the need to identify the secondary variables.

For many policy purposes, however, the effects of secondary variables are often critical, especially when the intervention is implemented as the result of a policy action. For this reason, the designs of RCFTs are often complex and incorporate individual differences among subjects and contextual variables so that their effects can be analyzed.

Even for the most rigorously conducted RCFTs, however, the results from one setting may not generalize to all other settings. Consequently, it may be difficult to identify "what works" in different settings from just one RCFT. Moreover, the use of RCFTs may be limited because they often require much time and expense in comparison with other approaches, or they may be precluded by ethical considerations.

Still, myriad RCFTs have been successfully conducted to inform social policy. *The Digest of Social Experiments* (Greenberg and Shroder, 2004) and its successor journal, *Randomized Social Experiments*, provide many examples.

OBSERVATIONAL STUDIES

Observational studies are nonexperimental research studies in which subjects or outcomes are observed and measured. If two groups are to be compared, the assignment of subjects among the two groups is not under the direct control of the investigator. Two types of observational studies are *quasi-experiments* (Campbell and Stanley, 1963) and *natural experiments* (see, e.g., Campbell and Ross, 1968). In the former, the investigator may manipulate the intervention; in the latter, it arises naturally. In neither type of study, however, does the investigator control which subjects receive the treatment. Observational studies can be more than passively observing data and analyzing them: for example, they may involve systematic measurement and aspects of "control," such as manipulating the timing of an intervention to predefined although nonrandomized groups.

Because they do not involve randomization, however, observational studies may not control for the effects of secondary variables. Without experimental confirmations, the observed outcomes could be the result of any combination of a range of confounding factors. For example, subjects may be self-selected, such as students in a private school who are to be compared with students in a public school, or they may be selected by others but with different characteristics, known or unknown, that may influence the outcome of the intervention. This possible influence is called *selection bias*. If there is selection bias, how the intervention affects the outcome for the treatment group in comparison with the control group must be described by a model, and that model will always include some assumptions. The model may or may not help with inference for what would have happened in a randomized experiment (see National Research Council, 1998). Moreover, the assumptions underlying the model may not be widely accepted in the scientific community.

Observational studies, however, are important in revealing important associations and in guiding the formulation of theory and models. The observation of a single case can reveal unsuspected patterns and provide explanations for unmotivated forms of behavior. As put by Coburn et al. (2009, p. 1,121): "The in-depth observation made possible by the single case study provides the opportunity to generate new hypotheses or build theory about sets of relationships that would otherwise have remained invisible."

Observational studies also serve many other important purposes for the use of social science knowledge as evidence for public policy. The country's wide range of longitudinal studies, for example, provides much information to guide public policy, from the extent to which people save for retirement (information provided by the Health and Retirement Study) to what different types of social welfare program benefits are actually obtained by families living in poverty (information from the Survey of Income and Program Participation). Observational studies, together with historical studies, provide the rich context in which public policy can benefit society. This use may be their most important role.

EVALUATION

Policies are typically implemented with large and highly heterogeneous populations. Even if a policy is based on carefully designed RCFTs or other studies, implementation beyond the confines of the original study population requires careful monitoring and evaluation to make sure that the results observed in the study hold in a larger context.

A researcher must always ask if the new program is producing similar desirable outcomes in the general population as it did in the experimental setting. In the absence of a closely monitored implementation program, issues of measurement, interpretation, and purposeful or accidental deviations from a protocol inevitably creep in, with unpredictable effects on the outcome. When policies are implemented in the general population, it may be done without carefully planned designs and randomized allocation of units to treatments. Unless close monitoring of the policy occurred during implementation, it may not even be known whether the intervention as it was originally devised was what was actually implemented.

Furthermore, the ultimate goal of a policy intervention may well be something to be observed in the future, when follow-up data may be difficult to obtain. For example, although some intermediate outcomes of a program to integrate addicts into the labor force—such as the proportion of participants who are drug free and are employed after a month of treatment—can be measured more or less precisely, it is much more difficult to determine that proportion a year after treatment. Moreover, even if one is able to obtain those data, how could one determine that the results are attributable to the program and not to other factors?

Today's trend toward accountability means that anyone proposing a new policy or intervention is also expected to prove that the intervention will "work." Thus, thinking about credible approaches to carry out evaluation studies is almost as critical as conducting the study itself. The principles of experimental design can play an important role, even for observational evaluation.

One approach, for example, is to compare a population before and after an intervention has occurred. As long as the study includes a well-defined reference group and as long as the investigator is reasonably certain that selection bias is not important, such studies can offer some evidence of the effectiveness (or lack thereof) of an intervention. Alternatively, an evaluation study can be planned as an RCFT, in which the goal is to understand whether the original conclusions about the efficacy of the intervention hold when other factors (e.g., the target population) are not exactly the same.

Both experimental and observational studies can be used to evaluate the long-term effects of interventions. An example of such an experimental study is the work of Kellam et al. (2008) on the effect on behavioral, psychiatric, and social outcomes in young adults of a classroom behavior management program carried out when they were in first and second grades. An example of an observational study is the work of Goodman et al. (2012) on the effects of childhood physical and mental problems on adult life, based on an analysis of longitudinal data from the British National Child Development Study.

The evaluation and monitoring of an intervention as implemented is closely related to the more general concept of *evolutionary learning*, a process to explore how the outcome of interest responds to changes in the original intervention. Consider, for example, a new teaching method shown to be effective in a small class setting. Will it also be as effective when class sizes are large?

A critical aspect of evolutionary learning is the need to proceed in a highly controlled manner in order to understand which factor or which combination of several factors that can be varied are influencing the outcome. Alternatively, a sequence of experiments can be designed in which two or more factors are varied according to a specified plan. In the absence of carefully designed sequential learning studies, it may be difficult to untangle the effect on the outcome of each of several factors under investigation.

As in the case of evaluation and monitoring, there is a theoretical framework developed for sequential learning in studies in which the response of interest is an unknown and may be a complex function of a large number of inputs. The approach is often known as response surface analysis: it was developed for engineering processes in the early 1950s by Box and Wilson (1951). The idea is to sequentially vary the settings of the input variables so that the response keeps improving.

Although developed for engineering processes, where it is known as evolutionary operation (Box and Draper, 1969), the approach appears to be well suited for the social sciences, in which the relationship between inputs and outputs is typically difficult to measure precisely (see the discussion in Fienberg et al., 1985). It is akin to what is referred to as a *learning system* that takes full advantage of each application of an intervention and extends the opportunity for discovery throughout the life-cycle of the intervention: its development, implementation, and evaluation.

META-ANALYSIS

Meta-analysis is an application of quantitative methods to combine the results of different studies (see Wachter and Straf, 1990). In such an analysis, a statistical analysis is typically made of a common numerical summary, such as an effect size, drawn from different studies (Hedges and Olkin, 1985). Today, there are many guides to conducting a meta-analysis: see, for example, Cooper (2010) and Cooper et al. (2009). Meta-analyses can lead to new hypotheses and theories and inform the design of an experiment or other research study to test them.

A major purpose of meta-analyses and other research syntheses is to reduce the uncertainty of cause-and-effect assessments of policy or program interventions. By statistically combining the results of multiple experiments, for example, the effect of a policy or program can be estimated more precisely than from any single study of an intervention. Moreover, comparing studies that are conducted with different participants in different settings allows for the examination of how different contexts affect the outcomes of a policy or program. However, if individual studies are flawed, then so will be a meta-analysis of them: thus, meta-analyses often specify standards of quality for the studies to be included.

The amalgamation of results from disparate studies can also be done with careful statistical modeling that is distinct from the approaches of meta-analysis. A good example of this approach is *Toxicological Effects of Methylmercury* (National Research Council, 2000b): its analysis is based on Bayesian methods developed by Dominici et al. (1999) to pool dose-response information across a relatively large number of studies. Other examples are in Neuenschwander et al. (2010) and Turner et al. (2009).

Work on understanding how to evaluate effectiveness of a policy intervention from the total body of relevant research assembled from inter-disciplinary studies has not been fully developed. An example of success, however, is researchers in early childhood intervention who have integrated knowledge about the developing brain, the human genome, molecular biology, and the interdependence of cognitive, social, and emotional development. These researchers have built a unified science-based framework for guiding priorities for early childhood policies around common concepts from neuroscience and developmental-behavioral research and broadly accepted empirical findings from four decades of program evaluation studies: see, for example, Center on the Developing Child at Harvard University (2007).

QUALITATIVE RESEARCH

In addition to experimental and observational studies, *qualitative research* can play important roles in developing knowledge about the societal consequences of a policy. The term covers many different types of studies, including ethnographic, historical, and other case studies; focus group interviews; content analysis of documents; interpretive sociology; and comparative and cross-national studies. The research may be derived from documentary sources, field observations, interviews with individuals or groups, and discourse between participants and researchers.

Structured, focused case comparisons are an important example of qualitative research. They are particularly useful when it is difficult to carry out studies that require high levels of control (see George, 1979; George and Bennett, 2005). By compiling and comparing case studies, it is possible to refine theory and also to develop useful assessments of the effectiveness of various types of policy interventions and the conditions that favor the effectiveness of one or another policy strategy. Structured case comparison methods have been used to inform diplomacy (Stern and Druckman, 2000) and assess policy strategies for resolving international conflicts (National Research Council, 2000a), to manage environmental resources at levels from local to global (National Research Council, 2002; Ostrom, 1990), and to evaluate efforts to engage the public in environmental decisions (Beierle and Cayford, 2002; National Research Council, 2008).

Archival studies are another example of qualitative research. They involve applying a model based on past evidence or decisions to a behavior or intervention for purposes of predicting future behavior (see, e.g., Institute of Medicine, 2010). Archival data may include public data sets collected by academic institutions or government agencies, such as Supreme Court records and corporate filings, or private data sets, such as medical records collected by public or private organizations.

Qualitative research allows for a rich assessment of respondents, often unattainable in other types of studies (Institute of Medicine, 2010). Like some quantitative observational studies, they can provide the rich context in which public policy can benefit society.

THE FUTURE: NEW SOURCES OF DATA

Advances in social science and in computing technology have generated a wealth and diversity of data sources. Although privacy and proprietary concerns pose ongoing challenges for the accessibility of these sources to researchers, the data represent tremendous potential and opportunities to study social phenomena in unprecedented ways.

Federal, state, and local governments collect administrative data on populations as a by-product of program responsibilities, such as the employment history data maintained by the Social Security Administration and the personal income and wealth data maintained by the Internal Revenue Service. There are health records, school records, land-use records, and much more. A growing interest in improving and using administrative records for scientific and policy purposes has generated increased attention to the issues of privacy, data sharing, data quality, and representativeness that have been central to census and survey data for decades.

The challenges and opportunities are even more pronounced with regard to digital data. With the rise and diffusion of advanced information, communication, and computing technologies, an astounding quantity of electronic data—from demographic and geographic variables to transaction records—is amassed at an exponential rate (see Prewitt, 2010). Though much of it is commercially collected and thus proprietary, the vast reservoir of digital data increasingly includes data collected by government agencies for public use. With respect to data quality, use is constrained by the relative brevity of the time series available for variables for which collection began only recently, as well as the fact that the definitions of variables are constantly changing.

The sheer quantity and diversity of digital data with the potential for social scientific use is astounding. As examples, consider continuous-time location data from cell phones; health data from electronic medical records and monitoring devices; consumer data from credit card transactions, online product searches and purchases, and product radio-frequency identification; satellite imagery and other forms of geocoded data; and data from social networking and other forms of social media.

The increasing "democratization of data" will enable policy analysts and policy makers to obtain much information for themselves, and it will continue to open new frontiers for social scientists. Automated information extraction and text mining have the potential to extract relevant data from the unstructured text of emails, social media posts, speeches, government reports, product reviews, and other web content. Crowd sourcing can be done through extracting information from social network websites. Longitudinal data can be compiled on millions of people with information on their locations, financial transactions, and communications. The data can be analyzed with methods of the emerging field of computational social science: network analysis, geospatial analysis, complexity models, and system dynamics, agent-based, and other social simulation models. Researchers and interested policy actors have only begun to scratch the surface of the potential of new data sources to contribute to policy making (King, 2011).

REFERENCES

Beierle, T.C., and Cayford, J. (2002). *Democracy in Practice: Public Participation in Environmental Decisions.* Washington, DC: Resources for the Future.

Box, G.E.P., and Draper, N.R. (1969). *Evolutionary Operation: A Statistical Method for Process Improvement.* New York: Wiley.

Box, G.E.P., and Wilson, K.B. (1951). On the experimental attainment of optimum conditions (with discussion). *Journal of the Royal Statistical Society, Series B, 13*(1), 1-45.

Campbell, D.T., and Ross, H.L. (1968). The Connecticut crackdown on speeding: Time-series data in quasi-experimental analysis. *Law and Society Review, 3*(1), 33-54.

Campbell, D.T., and Stanley, J.C. (1963). *Experimental and Quasi-Experimental Designs for Research.* Boston, MA: Houghton Mifflin.

Center on the Developing Child at Harvard University. (2007). *A Science-Based Framework for Early Childhood Policy: Using Evidence to Improve Outcomes in Learning, Behavior, and Health for Vulnerable Children.* Available: http://developingchild.harvard.edu/files/7612/5020/4152/Policy_Framework.pdf [August 2012].

Coburn, C.E., Toure, J., and Yamashita, M. (2009). Evidence, interpretation, and persuasion: Instructional decision making at the district central office. *Teachers College Record, 111*(4), 1,115-1,161.

Cooper, H.M. (2010). *Research Synthesis and Meta-Analysis: A Step-by-Step Approach* (fourth edition). Thousand Oaks, CA: Sage.

Cooper, H.M., Hedges, L.V., and Valentine, J.C. (Eds.). (2009). *The Handbook of Research Synthesis* (second edition). New York: Russell Sage Foundation.

Dominici, F., Zeger, S.L., and Samet, J.M. (1999). Combining evidence on air pollution and daily mortality from the largest 20 U.S. cities: A hierarchical modeling strategy (with discussion). *Journal of the Royal Statistical Society, Series A, 163,* 263-302.

Fienberg, S.E., Singer, B., and Tanur, J. (1985). Large-scale social experimentation in the United States. In A.C. Atkinson and S.E. Fienberg (Eds.), *A Celebration of Statistics: The ISI Centenary Volume* (pp. 287-326). New York: Springer Verlag.

George, A.L. (1979). Case studies and theory development: The method of structured, focused comparison. In P.G. Lauren (Ed.), *Diplomacy: New Approaches in History, Theory, and Policy.* New York: The Free Press.

George, A.L., and Bennett, A. (2005). *Case Studies and Theory Development in the Social Sciences.* Cambridge, MA: MIT Press.

Goodman, A., Joyce, R., and Smith, J.P. (2012). The long shadow cast by childhood physical and mental problems on adult life. *Proceedings of the National Academy of Sciences, 108,* 6,032-6,037.

Greenberg, D., and Shroder, M. (2004). *The Digest of Social Experiments* (third edition). Washington, DC: Urban Institute Press.

Hedges, L.V., and Olkin, I. (1985). *Statistical Methods for Meta-Analysis.* San Diego, CA: Academic Press.

Institute of Medicine. (2010). *Bridging the Evidence Gap in Obesity Prevention: A Framework to Inform Decision Making.* Committee on an Evidence Framework for Obesity Prevention Decision Making, S.K. Kumanyika, L. Parker, and L.J. Sim, Eds. Food and Nutrition Board. Washington, DC: The National Academies Press.

Kellam, S.G., Reid, J., and Balster, R.L. (2008). Effects of a universal classroom behavior program in first and second grades on young adult problem outcomes, *Drug and Alcohol Dependence, 95,* S1-S4.

King, G. (2011). Ensuring the data-rich future of the social sciences. *Science, 331*(6,018), 719-721.

National Research Council. (1998). *Assessing Evaluation Studies: The Case of Bilingual Education Strategies.* Panel to Review Evaluation Studies of Bilingual Education, M.M. Meyer and S.E. Fienberg, Eds. Committee on National Statistics, Commission on Behavioral and Social Sciences and Education. Washington, DC: National Academy Press.

National Research Council. (2000a). *International Conflict Resolution After the Cold War.* Committee on International Conflict Resolution, P.C. Stern and D. Druckman, Eds. Washington, DC: National Academy Press.

National Research Council. (2000b). *Toxicological Effects of Methylmercury.* Committee on the Toxicological Effects of Methylmercury, Board on Environmental Studies and Toxicology, Commission on Life Sciences. Washington, DC: National Academy Press.

National Research Council. (2002). *The Drama of the Commons.* Committee on the Human Dimensions of Global Change, E. Ostrom, T. Dietz, N. Dolsak, P.C. Stern, S. Stonich, and E.U. Weber, Eds. Committee on the Human Dimensions of Global Change, Division of Behavioral and Social Sciences and Education. Washington, DC: The National Academies Press.

National Research Council. (2008). *Public Participation in Environmental Assessment and Decision Making*. Panel on Public Participation in Environmental Assessment and Decision Making, T. Dietz and P.C. Stern, Eds. Committee on the Human Dimensions of Global Change, Division of Behavioral and Social Sciences and Education. Washington, DC: The National Academies Press.

Neuenschwander, B., Capkun-Niggli, G., Branson, M., and Spiegelhalter, D.J. (2010). Summarizing historical information on controls in clinical trials. *Clinical Trials*, 7(1), 5-18.

Ostrom, E. (1990). *Governing the Commons: The Evolution of Institutions for Collective Action*. New York: Cambridge University Press.

Prewitt, K. (2010). Science starts not after measurement but with measurement. *The Annals of the American Academy of Political and Social Sciences*, 631(1), 7-13.

Stern, P.C., and Druckman, D. (2000). Evaluating interventions in history: The case of international conflict resolution. *International Studies Review*, 2(1), 33-63.

Turner, R.M., Spiegelhalter, D.J., Smith, G.C.S., and Thompson, S.G. (2009). Bias modelling in evidence synthesis. *Journal of the Royal Statistical Society, Series A, 172*, 23-47.

Wachter, K.W., and Straf, M.L. (1990). *The Future of Meta-Analysis*. New York: Russell Sage Foundation.

Appendix B

Biographical Sketches of Committee Members and Staff

Kenneth Prewitt (*Chair*) is the Carnegie professor of public affairs at Columbia University. Previously, he taught at the University of Chicago, Stanford University, Washington University, in Kenya and Uganda. His other positions included director of the U.S. Census Bureau and of the National Opinion Research Center, president of the Social Science Research Council, senior vice president of the Rockefeller Foundation, and dean at the New School University. His current writing focuses on how to improve race statistics and why that matters and the use of science in policy interests. He is a fellow of numerous professional associations and broadly active in science policy. He has a Ph.D. in political science from Stanford University.

George W. Bohrnstedt is senior vice president for research (*emeritus*) at the American Institutes for Research, where he helped in the development of new programs of research for the organization. He has had an interest in measurement in the social sciences throughout his professional career, growing out of his minor in educational psychology with an emphasis on tests and measurement. He currently chairs the National Center for Education Statistics' Validity Studies Panel for the National Assessment of Educational Progress (NAEP) and works on two other NAEP research projects. He has B.S., M.S., and Ph.D. degrees in sociology and a minor in educational psychology from the University of Wisconsin–Madison.

Norman M. Bradburn is Tiffany and Margaret Blake distinguished service professor emeritus of the University of Chicago and a senior fellow at the National Opinion Research Center (NORC). Associated with NORC since 1961, he has been both director and president of its Board of Trustees. At the National Research Council, he has chaired the Committee on National Statistics, the panel to advise the Census Bureau on alternative methods for conducting the census in the year 2000, the panel to review the National Assessment of Educational Progress, and the panel to assess the 2000 census. From 2000-2004, he was the assistant director for the Social, Behavioral, and Economic Sciences Directorate at the National Science Foundation. Bradburn has a Ph.D. in social psychology from Harvard University.

Alicia L. Carriquiry is distinguished professor of statistics at Iowa State University. Her research interests are in Bayesian statistics and general methods. Her recent work focuses on nutrition and dietary assessment, as well as on problems in genomics, forensic sciences, and traffic safety. Carriquiry is an elected member of the International Statistical Institute, a fellow of the Institute of Mathematical Statistics, and a fellow of the American Statistical Association. She has served on the executive committee of the Institute of Mathematical Statistics, of the International Society for Bayesian Analysis, and of the American Statistical Association and was a member of the board of trustees of the National Institute of Statistical Sciences. She has served on several committees and panels of the National Academies including the standing Committee on National Statistics. She is currently chairing a committee that is discussing approaches to estimate the number of illegal border crossings in the Southwestern border of the United States. She has a M.Sc. in animal science from the University of Illinois at Urbana-Champaign, and a M.Sc. in statistics and a Ph.D. in statistics and animal genetics from Iowa State University.

Nancy D. Cartwright is professor of philosophy in the Department of Philosophy, Logic and Scientific Method in the London School of Economics and Political Science; she is also professor of philosophy at the University of California, San Diego. Her principal interests are the philosophy and history of science (especially physics and economics), causal inference, and evidence and objectivity in science and policy. She has recently served as president of the Philosophy of Science Association and of the American Philosophical Association, Pacific Division. Cartwright has a Ph.D. in philosophy from the University of Illinois at Chicago.

Harris Cooper is professor of psychology and chair of the Department of Psychology and Neuroscience at Duke University. His work involves research syntheses and meta-analysis in varied fields, such as personality and social psychology, developmental psychology, marketing, and developmental medicine and child neurology; he is also interested in the application of social and developmental psychology to education policy issues. He is past editor of the *Psychological Bulletin* and currently serves as the chief editorial adviser for the journals program of the American Psychological Association. He has a Ph.D. in social psychology from the University of Connecticut.

Jonathan R. Dolle is a research associate for evaluation and field building at the Carnegie Foundation for the Advancement of Teaching, where he also directs the foundation's postbaccalaureate fellowship program. His current work focuses on understanding how education organizations can adapt tools and methods from quality improvement efforts in health care and manufacturing. From 2005 to 2010, Dolle worked as a research assistant on Carnegie's business education and liberal learning project, where he co-authored the book *Rethinking Undergraduate Business Education.* In the fall of 2009, he was a Mirzayan policy fellow at the National Academy of Sciences. He has a Ph.D. in education from the Stanford University School of Education and degrees in engineering, philosophy, and education policy from the University of Illinois at Urbana-Champaign.

Michael J. Farrell was appointed deputy commissioner for strategic initiatives in the New York City Police Department in January 2002. In this position, he directs the activities of the Office of Management Analysis and Planning and the Quality Assurance Division. He was first appointed to the New York City Police Department in 1985 as the director of special projects and has since served as assistant commissioner, Office of the First Deputy Commissioner; deputy commissioner for policy development; and as deputy commissioner for policy and planning. From June 1999 to January 2002, he served as the deputy director of criminal justice for New York state, providing oversight and coordination of the state's criminal justice agencies. Prior to his tenure with the New York City Police Department, he served on the director's staff at the National Institute of Justice, the research branch of the U.S. Department of Justice.

Stephen E. Fienberg is Maurice Falk university professor of statistics and social science in the Department of Statistics, the Machine Learning

Department, the Heinz College, and Cylab at Carnegie Mellon University. A leader in the development of statistical methods for the analysis of multivariate categorical data, he has also worked on the development of statistical methods for large-scale sample surveys and censuses, such as those carried out by the federal government, and on the interrelationships between sample surveys and randomized experiments. His current research includes technical and policy aspects of privacy and confidentiality and on methods for the analysis of network data. Fienberg has also been active in the application of statistical methods to legal problems and in assessing the appropriateness of statistical testimony in actual legal cases, and he has linked his interests in Bayesian decision making to the issues of legal decision making. He has served on a broad array of National Research Council and Institute of Medicine committees, evaluating scientific evidence arising from the social, behavioral, and biomedical science studies. He is a member of the National Academy of Sciences and a fellow of the American Academy of Arts and Sciences and the Royal Society of Canada. Fienberg has a Ph.D. in statistics from Harvard University.

Sheila S. Jasanoff is Pforzheimer professor of science and technology studies at Harvard University's John F. Kennedy School of Government, where she directs the Program on Science, Technology and Society. Her research focuses on the relationship of science and technology to law, politics, and policy in modern democratic societies, with particular emphasis on the role of science in cultures of public participation and public reasoning. She has written and lectured widely on environmental regulation, risk management, and the politics of the life sciences in the United States, Europe, and India. She has a Ph.D. in linguistics from Harvard University and a J.D. from Harvard Law School.

Robert L. Jervis is Adlai E. Stevenson professor of international politics at Columbia University, where he has been a member of the faculty since 1980. He has also taught at the University of California, Los Angeles (1974-1980) and Harvard University (1968-1974). In 2000-2001, he served as the president of the American Political Science Association. Jervis is co-editor of *Studies in Security Affairs* and a member of numerous editorial review boards for scholarly journals. Most recently, his publications include *Why Intelligence Fails* (2010), as well as edited volumes and numerous articles in scholarly journals. He has a Ph.D. from the University of California, Berkeley.

Robert E. Litan is director of research for Bloomberg Government. He was previously vice president for research and policy at the Kauffman Foundation, where he managed and conducted research relating to entrepreneurship, and a senior fellow in the Economic Studies Program at the Brookings Institution. He is the co-author of *Better Capitalism* (Yale University Press, 2012), *Good Capitalism, Bad Capitalism, Economic Growth and Prosperity* (2007), and *Competitive Equity: Developing a Lower Cost Alternative for Mutual Funds* (2007). Litan has served on the staff of the Council of Economic Advisers, as deputy assistant attorney general in the Antitrust Division of the Justice Department, and as an associate director of the Office and Management and Budget. He also has been a consultant to the Treasury Department on financial policy issues. He was a member of the Commission on the Causes of the Savings and Loan Crisis. Litan has a B.S. degree in economics (summa cum laude) from the Wharton School Department of Finance at the University of Pennsylvania; a J.D. from Yale Law School; and a Master of Philosophy and a Ph.D. in economics from Yale University.

Ann Morning is associate professor of sociology at New York University. Morning publishes and lectures on racial classification and conceptualization in the United States and abroad, with particular attention to the uses of racial categorization in demography, law, medicine, and genetic research. Her research topics include the historical and contemporary demography of the U.S. multiracial population, racial classification of ethnic groups like Hispanic and South Asian Americans, cross-national comparison of ethnic classification practices on censuses worldwide, scientific and lay concepts of race, and the effect of socially desirable reporting on Americans' expression of biological definitions of race. She has a B.A. (magna cum laude) in economics and political science from Yale University and a Master of International Affairs from Columbia University's School of International and Public Affairs. She also has an M.A. and Ph.D. in sociology from Princeton University, where she specialized in demography at the Office of Population Research. Her doctoral dissertation won the American Sociological Association's Dissertation Award in 2005, and was published in 2011 by the University of California Press as *The Nature of Race: How Scientists Think and Teach about Human Difference.*

Robert A. Pollak is Hernreich distinguished professor of economics in the Faculty of Arts and Sciences and the John M. Olin School of Business at Washington University in St. Louis. His research interests include the economics of the family, price and cost-of-living indexes, and environmental policy. At the National Research Council, he served on the Committee on National Statistics panel on cost-of-living indexes. From 1997 to 2007, Pollak co-chaired the MacArthur Foundation Network on the Family and the Economy, an interdisciplinary group of economists, sociologists, and developmental psychologists studying the functioning of families. He has a Ph.D. in economics from the Massachusetts Institute of Technology.

Melissa Lee Sands is a Ph.D. student in government at Harvard University, where she studies American politics and quantitative methodology. She holds a Master of Public Administration from Columbia University's School of International and Public Affairs, where she concentrated in advanced policy and economic analysis, and a B.A. in sociology from the University of Wisconsin–Madison. She has held an associate faculty appointment at SIPA and has worked for public officials in Wisconsin and for nonprofit organizations in Madison, Wisconsin; New York City; and Lagos, Nigeria.

Stephen H. Schneider (deceased July 2010) was the Melvin and Joan Lane professor for interdisciplinary environmental studies, professor in the Department of Biology, and a senior fellow in the Woods Institute for the Environment at Stanford University. He was also a professor by courtesy in the Department of Civil and Environmental Engineering. He served as a research scientist at the National Center for Atmospheric Research from 1973 to 1996, where he co-founded the Climate Project. He focused on climate change science, integrated assessment of ecological and economic impacts of human-induced climate change, and identifying viable climate policies and technological solutions. He consulted for federal agencies and White House staff in six administrations. Involved with the Intergovernmental Panel on Climate Change (IPCC) since 1988, he was coordinating lead author of Working Group II for Chapter 19, "Assessing Key Vulnerabilities and the Risk from Climate Change," and a core writer for the Fourth Assessment Synthesis Report. He, along with four generations of IPCC authors, received a collective Nobel Peace Prize in 2007. Elected to the National Academy of Sciences in 2002, Schneider received the American Association for the Advancement of Science/Westinghouse Award for Public Understanding of Science and Technology and a MacArthur Fellowship for integrating and interpreting the results of global climate research. Founder and editor of *Climatic Change*, he authored or co-authored many books, scientific papers, proceedings, legislative testimonies, edited books and chapters, reviews, and editorials.

Thomas A. Schwandt is professor in the Department of Educational Psychology at the University of Illinois at Urbana-Champaign. Previously he was a faculty member in the School of Education at Indiana University, where he was also a fellow in the university's Poynter Center for the Study of Ethics and American Institutions. He has also held a faculty appointment in medical education at the University of Illinois at Chicago Medical School. He is the author of *Evaluation Practice Reconsidered* (2004) and *The Dictionary of Qualitative Inquiry* (1997, 2001, 2007), among others. In addition, he has authored many papers and chapters on issues in the theory of evaluation and interpretive methodologies. In 2002, he received the Paul F. Lazarsfeld Award from the American Evaluation Association for his contributions to evaluation theory. Schwandt has a Ph.D. in inquiry methodology from Indiana University, Bloomington.

Miron L. Straf (*Study Director*) is deputy director of the Division of Behavioral and Social Sciences and Education at the National Research Council. Previously, he served as director of the division's Committee on National Statistics and was at the National Science Foundation, where he worked on developing the research priority area for the social, behavioral, and economic sciences. He was on the faculty of the University of California, Berkeley, and the London School of Economics and Political Science, and was president of the American Statistical Association. He received the American Association of Public Opinion Research's Innovators Award for his work on cognitive aspects of survey methodology. His major research interests are government statistics and the use of statistics and research for public policy decision making. He has a Ph.D. in statistics from the University of Chicago.

Sidney Verba is Carl H. Pforzheimer university professor emeritus in the Department of Government at Harvard University and director emeritus of the Harvard University Library. He is a member of the National Academy of Sciences, a fellow of the American Academy of Arts and Sciences and the American Philosophical Society, and president emeritus of the American Political Science Association (APSA). He has received numerous APSA awards, including the Krammerer Prize, the Woodrow Wilson Prize, and the James Madison Prize, APSA's highest prize awarded every 3 years for a career contribution to political science. In 2002, he was awarded the Johan Skytte Prize, the major international prize in political science. He received a Ph.D. from Princeton University in 1959.